Solar Thermal and Biomass Energy

WITPRESS

WIT Press publishes leading books in Science and Technology.
Visit our website for the current list of titles.
www.witpress.com

WITeLibrary

Home of the Transactions of the Wessex Institute, the WIT electronic-library provides the international scientific community with immediate and permanent access to individual papers presented at WIT conferences. Visit the WIT eLibrary at
http://library.witpress.com

Solar Thermal and Biomass Energy

G. Lorenzini, C. Biserni & G. Flacco

University of Bologna, Italy

WIT*PRESS* Southampton, Boston

G. Lorenzini, C. Biserni & G. Flacco
University of Bologna, Italy

Published by

WIT Press
Ashurst Lodge, Ashurst, Southampton, SO40 7AA, UK
Tel: 44 (0) 238 029 3223; Fax: 44 (0) 238 029 2853
E-Mail: witpress@witpress.com
http://www.witpress.com

For USA, Canada and Mexico

WIT Press
25 Bridge Street, Billerica, MA 01821, USA
Tel: 978 667 5841; Fax: 978 667 7582
E-Mail: infousa@witpress.com
http://www.witpress.com

British Library Cataloguing-in-Publication Data

A Catalogue record for this book is available
from the British Library

ISBN: 978-1-84564-147-4

Library of Congress Catalog Card Number: 2008909438

*The texts of the papers in this volume were set
individually by the authors or under their supervision.*

No responsibility is assumed by the Publisher, the Editors and Authors for any injury and/or damage to persons or property as a matter of products liability, negligence or otherwise, or from any use or operation of any methods, products, instructions or ideas contained in the material herein. The Publisher does not necessarily endorse the ideas held, or views expressed by the Editors or Authors of the material contained in its publications.

© WIT Press 2010

Printed in Great Britain by the MPG Books Group, Bodmin and King's Lynn

All rights reserved. No part of this publication may be reproduced, stored in a retrieval system, or transmitted in any form or by any means, electronic, mechanical, photocopying, recording, or otherwise, without the prior written permission of the Publisher.

Contents

Preface ix

PART I: SUN ENERGY

Chapter 1
The solar radiation 3

1. The solar physics 3
2. The solar constant 5
3. The extraterrestrial radiation 6
4. The position of the Sun in the celestial vault 6
5. The solar radiation on the Earth's soil during clear sky days 9
6. Instantaneous direct radiation received on a surface 12
7. Instantaneous global radiation received on a surface 15
8. Calculation of direct energy received on a surface 15
9. The true solar time 16
10. The diagram of solar trajectories 18
11. The monthly average solar radiation on inclined surfaces 19
12. Daily radiation on an inclined surface 20
13. Hourly solar radiation on inclined surfaces 20
14. The local radiation data retrieval 21
15. Variation in the energy which can be intercepted by the position of the surfaces 22

Chapter 2
Solar energy utilization 25

1. Introduction 25
2. Low-temperature solar thermal technology 26
   2.1 The advantages 26
   2.2 Low-temperature solar thermal system 27
       2.2.1 The collector 28
       2.2.2 Typologies of solar systems 54
       2.2.3 The solar circuit 69
       2.2.4 The storage tank 75
   2.3 Passive solar heating systems 80
       2.3.1 Direct gain systems 81
       2.3.2 Indirect gain systems 82

		2.3.3 Solar greenhouse	85
		2.3.4 Isolated gain systems	86
3	Medium-temperature solar thermal technology		87
4	High-temperature solar thermal technology		90
	4.1	Concentrating solar power technology: clean energy for power tenability	90
	4.2	Prospects of CSP technologies	93
	4.3	The Italian position and interest in CSP technologies	93
	4.4	CSP technology	94
		4.4.1 Linear parabolic collector systems	95
		4.4.2 Tower system with a central receiver	99
		4.4.3 Parabolic dish collector systems	104
		4.4.4 The use of CSP technology for electricity production	105
		4.4.5 The future: the direct production of solar hydrogen	106
	4.5	The ENEA technological proposal for solar electricity: the use of molten salts in parabolic collector systems	106
		4.5.1 The advantages of molten salts	108
		4.5.2 The solar collector used by ENEA	110
		4.5.3 The Archimedes Project	115
	4.6	Conclusions	118
	4.7	Solar technologies for electricity generation without light concentration	119
		4.7.1 Solar chimneys/towers	119
		4.7.2 Solar ponds	123

General bibliography and consulted websites – Part I **127**

PART II: BIOMASSES ENERGY

Chapter 3
Biomasses identities **133**

1	Introduction	133
2	Definition and classification	134
3	Origin and nature	135
	3.1 The forest and agro-forest behaviour	135
	3.2 The agricultural compartment	137
	3.2.1 Agricultural residuals	138
	3.2.2 Dedicated cultures	139
	3.3 The zoo technique compartment	143
	3.4 Industrial activities	145
	3.4.1 The wood industry	145
	3.4.2 The cellulose and paper industry	145
	3.4.3 The agro-alimentary industry	146
	3.5 Urban residuals	146
4	Commercial forms	147
	4.1 Liquid state combustible biomasses	147

	4.1.1 Firewood	147
	4.1.2 Chips	147
	4.1.3 Densified forms: pellets	150
	4.1.4 Densified forms: the briquette	153
	4.1.5 Lignocellulose biomass fitness for the transformation into commercial forms	155
4.2	Fuel biomass in the liquid state	156
	4.2.1 Vegetable oils	156
	4.2.2 Bio-diesel	156
	4.2.3 Bio-ethanol	159
4.3	Combustible biomasses in the gaseous state	162
	4.3.1 Bio-gas	162

Chapter 4
Energy from biomasses 167

1 Biomass energy conversion 167
2 Biochemical conversion 168
   2.1 Anaerobic digestion 169
      2.1.1 Plant typologies applicable to liquid or effluent manures 170
      2.1.2 Co-digestion 173
      2.1.3 Bio-gas in solid rejections dumps 174
   2.2 Aerobic digestion 175
   2.3 Alcohol fermentation 176
      2.3.1 The sacchariferous section of the bio-ethanol production spinneret 176
      2.3.2 The starchy section of the bio-ethanol production spinneret 177
      2.3.3 The cellulosic section of the bio-ethanol production spinneret 178
   2.4 Oil extraction and bio-diesel production 179
      2.4.1 Vegetable oil extraction 179
      2.4.2 Vegetable oil regeneration 180
      2.4.3 Transesterification 181
3 Thermochemical conversion 182
   3.1 Direct combustion 181
   3.2 Gasification 190
      3.2.1 Fixed-bed gasificators 191
      3.2.2 Fluid-bed gasificators 194
      3.2.3 Producer gas applications 197
   3.3 Pyrolysis 198

Chapter 5
Environmental aspects 203

1   Reduction of emissions into the atmosphere.. 203
   1.1   The carbon dioxide emissions balance... 203
   1.2   Comparison between the polluting emissions of
        the main vegetable and fossil origin fuels 204
       1.2.1   Bio-ethanol.. 204
       1.2.2   Bio-diesel.. 205
       1.2.3   Bio-gas.. 206

Bibliography and consulted websites – Part II 207

Index   209

Preface

Conventional energy sources based on oil, coal and natural gas are damaging economic and social progress, the environment and human life. Many people are concerned about these problems and wish to address the symptoms as a matter of urgency, but not all understand the basic causes and consequently do not realize that not only technological, but also social changes are required. It is now widely acknowledged that renewable energy capacity has to be increased by exploiting its enormous potential.

During the last few years the 'energy issue' has been assuming a more and more important role among any other choice, strategy and policy concerning human survival and development.

Nowadays the energy model is almost totally centred (for the 80%) on the exploitation of fossil fuels such as petrol, natural gas and coal. To the industrial–economic costs connected with these fuels, social and environment costs, which cannot be overlooked, have to be added.

First of all, fossil fuels are exhaustible energy sources; their formation time is infinitely lower than the one which refers to their exploitation and for this reason are also defined as 'non-renewable resources'. Although the level of the world's fossil fuel supply cannot be considered as worrying in the short term, the increased difficulties in reaching the fields have made the cost–benefit ratio of the extraction processes less and less favourable.

Secondly, the political, social and economic instability, deriving from the world consumption distribution (only 20% of world population consumes the 80% of available resources) and the increasing number of wars connected with the geopolitics of fossil resources and with the control of international supplies, represents a risk for the security and the possible normal development of nations.

Eventually it is necessary to consider the environmental impact caused by the exploitation of fossil energy sources; actually their combustion process brings on the emission of noxious substances such as sulphurous anhydride, nitrogen monoxide and carbon anhydride (15 billion tons of CO_2 are poured out into the atmosphere every year). Sulphurous anhydride and nitrogen monoxide contribute to the formation of acid rains while carbon anhydride is the main greenhouse gas which causes global warming (greenhouse effect). So behind fossil fuel exploitation is hidden the risk of worrying consequences regarding both the Earth (desertification, arctic ice melt, sea level rise...) and indirectly human health (rise in respiratory diseases, decrease of drinkable water...) .

The analyzed context brought us to review in a critical way the concepts and models of development which have been taken into consideration to date and which have centred on the massive exploitation of fossil sources. During the last few years this review has led to elaborate the concept of sustainable development, which is based on energy consumption reduction and optimization, and also on the use of renewable energy sources (the Sun, the Wind, hydraulic energy, geothermic resources, tides and wave motion; this definition is completed by the biomasses, although these resources can only be considered as renewable if run with the purpose to make their exploitation time consistent with their renewal time).

In comparison to fossil fuels, renewable sources could contribute to the development of a sustainable energy system and to environment and territorial protection; they could also provide new economic growth opportunities.

Recently, the European Union passed new legislative measures to delineate in a binding manner the plan, from now to 2020, to decrease the climate effects caused by present energy consumption levels; that is to say that at least 20% of primary energy will have to be produced by renewable sources, greenhouse gas emission will have to be reduced of 20% and another 20% will have to be an energy saving which the EU means to reach by a wide energy efficiency recovery.

The importance placed upon renewable energy sources now and in the future inside the world energy panorama led us to focus this study on what can be defined as the most relevant renewable source: the Sun.

A policy of energy sustainability can't leave solar energy exploitation out of consideration. Actually its incident quota on the terrestrial surface is 10,000 times greater than the yearly energy requirement of the world's population. Besides being the origin of almost all the other energy sources, renewable and conventional, excluding geothermic, nuclear and gravitational (tides) ones, the energy provided by the Sun is free, endless and clean (the devices used to exploit solar energy are characterized by very low emissions while running). Moreover solar energy is easy to harness and distribute (it is particularly abundant in many world areas with depressed and difficult economic situation).

The first chapter of this study is dedicated to the analysis and calculation of solar radiation incident on an inclined surface at an instantaneous, hourly and daily level.

The second chapter offers a summary and an analysis of all technologies available today to use solar energy: the solar thermal (technologies which exploit solar radiation in order to produce thermal energy that can be used in domestic, civil and productive fields; the differences between low, medium and high temperature solar thermal energy will be identified).

In the last part of the book we judge through a deeper investigation the opportunities offered by the exploitation of biomass energy.

Renewable energy education is a relatively new field and previously it formed a minor part of traditional university courses. However, over the past decade, several new approaches have emerged: we see these in the new literature and, even more clearly, in new books. The present treatise, in the authors' auspices, represents a contribution to this new 'incoming science'.

The book is highly recommended to professors, students and professionals in mechanical, civil, environmental, chemical and agricultural engineering. It is also recommended to all the readers interested in the aims, philosophy, structure, design, strategies and overcomes in the use of energy from 'solar thermal and biomasses'.

The Authors, 2010

PART I

Sun energy

CHAPTER 1

The solar radiation

1 The solar physics

The Sun is a sphere made up of gaseous elements consisting of 80% hydrogen, 19% helium and 1% of all the well-known substances. It has a diameter of $1.39 \cdot 10^9$ m and it is located at a distance of $1.495 \cdot 10^{11}$ m from the Earth. However, this distance may vary by $\pm 1.7\%$ during the year because of the orbit's eccentricity.

The Sun is characterized by two motions: a motion of revolution around the centre of the galaxy, which has a linear speed of 300 km/s and takes 200 million years to complete, and a motion of rotation around the axis, which lasts about 4 weeks.

Inside the Sun, numerous fusion reactions take place. The heat produced by these reactions spreads from the inner layers to the outer layers by convection, conduction and radiation. From the outer layers, the heat is transmitted to the surrounding space by radiation. Among the nuclear reactions that occur in the Sun, the most important is the one which converts hydrogen into helium; the mass of a helium nucleus is smaller than that of the four original protons and this mass defect is converted into energy.

The mass of the Sun is roughly $2 \cdot 10^{30}$ kg. The areas at the centre of the Sun reach temperatures of about 8–40 million kelvin and a density 100 times greater than that of water. However, the density is extremely lower in the outer layers.

It is believed that the region between 0 and 0.23R (R = solar ray), which constitutes 40% of the solar mass, produces 90% of the solar energy. The area between 0.7 and 1R is called the convective envelope (temperature 5000 K, density 10^{-5} kg/m^3), because of the importance of convective processes in this layer. The photosphere, the outer layer from the convective envelope, is composed of strongly ionized gases, which are capable of absorbing and emitting through a continuous spectrum of radiation. Over the photosphere, there is the inversion layer, which is hundreds of kilometres wide and is made up of cold gases. Outside the inversion layer, there is the chromosphere, which is 10,000 km wide, and the corona, characterized by a very low density and high temperatures (10^6 K) (Fig. 1) [1].

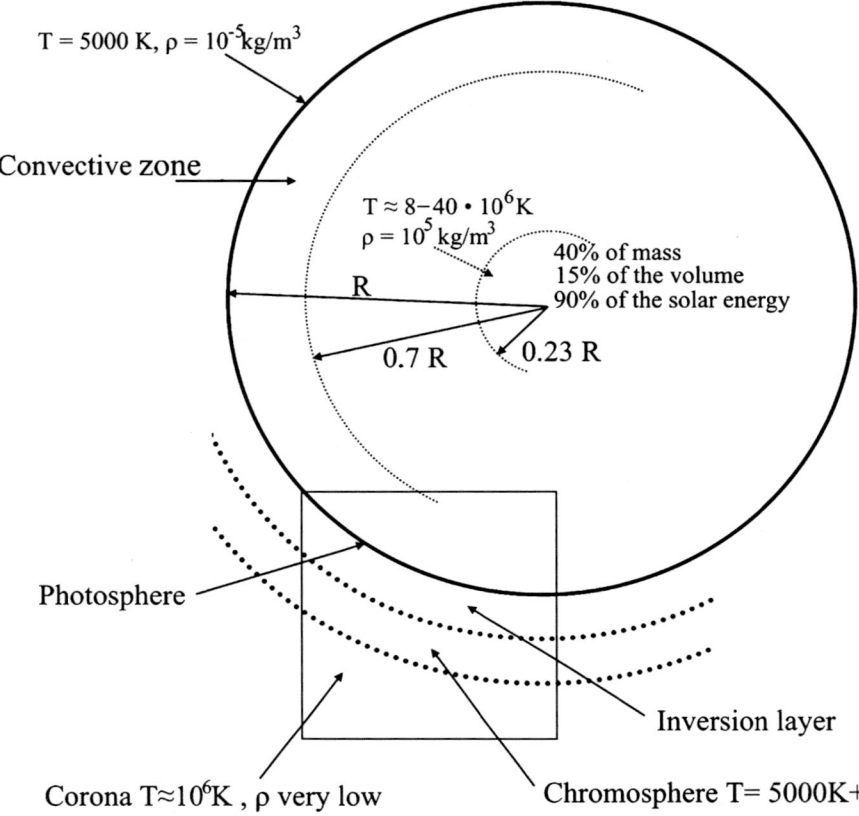

Figure 1: The Sun's structure.

The electromagnetic radiation emitted by the Sun extends over a wide wavelength interval: from 0.1 nm to 10^4 m; however, the greatest part of that energy falls in the interval between 0.2 and 4 μm. In particular, 95% of the energy which reaches the Earth is included between 0.3 and 2.4 μm. The spectrum of the solar radiation is similar to a black body's spectrum at a temperature of 5780 K, since temperatures at the surface of a star fluctuate between 4000 and 6000 K. Therefore, it is right to assume that the behaviour of the Sun with regard to radiation is similar to the behaviour of a black body at a uniform/regular temperature (Fig. 2). This temperature of 5780 K is calculated using the Stefan–Boltzmann law [1, 3].

Analysing the spectrum more carefully, one can notice that the greatest part of the radiation falls in (1) the ultraviolet band, which extends from 0.20 to 0.38 μm; (2) the visible light band, from 0.38 to 0.78 μm; and (3) the near infrared band until about 4 μm. Only 8–9% of all the solar energy which reaches the Earth falls in the ultraviolet band; 46–47% falls in the visible band while the remaining 45% falls in the infrared band [3].

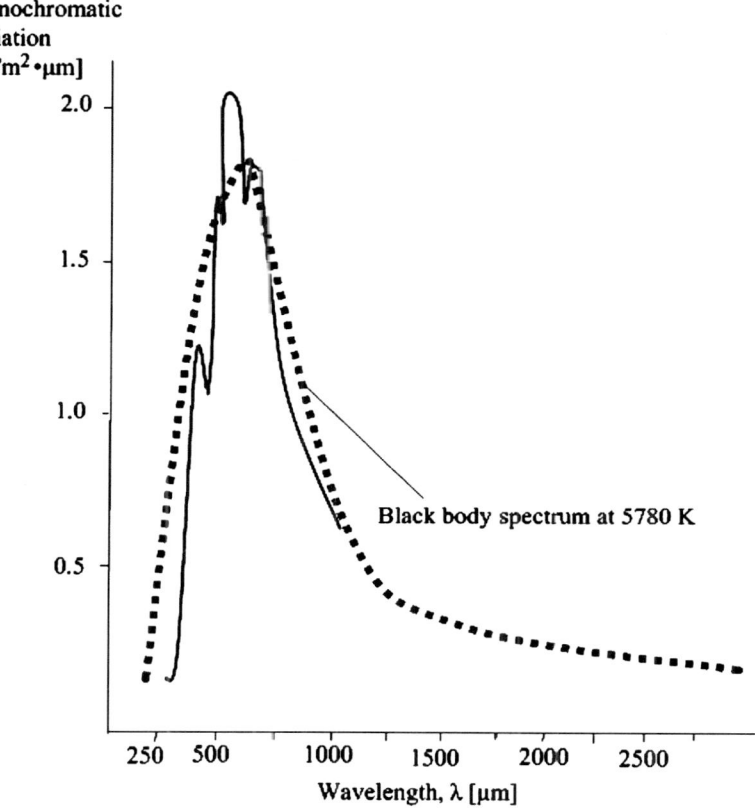

Figure 2: Monochromatic radiation outside the Earth's atmosphere.

2 The solar constant

The *solar constant* I_{cs} is the average energy radiated by the Sun per time unit on a unitary surface situated outside the Earth's atmosphere and perpendicular to the Sun's rays. It measures 1367 W/m². Considering the atmospheric phenomenon of absorption and diffusion and the Sun's inclination above the horizon, on the Earth's soil the solar constant reaches a maximum of 1000 W/m² (radiation on land, at midday, during a clear sky day) [1, 4].

The total power radiated by the Sun can be calculated as follows:

$$P = 4\pi R_m^2 I_{cs} = 4\pi(150 \cdot 10^9)^2 1367 = 3.8 \cdot 10^{26} \text{ W} \quad (1)$$

where R_m is the average distance between the Earth and the Sun [2].

The Earth intercepts only $1.73 \cdot 10^{17}$ W of that power. Owing to nuclear reactions, a mass of $4.27 \cdot 10^9$ kg can be destroyed in a second; thus, nearly 0.0067% of the solar mass will be lost in a billion years [1].

Once we know the power emitted by the Sun, it is easy to calculate the heat produced internally per unit of solar volume:

$$q = P/(4/3)\pi R^3 = 3.8 \cdot 10^{26} /(4/3)\pi(7.25 \cdot 10^8)^3 = 0.24 \text{ W}/\text{m}^3 \qquad (2)$$

where $R (= 7.25 \cdot 10^8$ m) is the solar ray.

The quantity calculated above is a particularly low value considering that, for example, the human body's heat production per unit volume is roughly 1400 W/m³ [2].

3 The extraterrestrial radiation

Since the Earth's orbit around the Sun is elliptical, the distance between them varies during the year, causing a ±3.3% fluctuation of the extraterrestrial radiation (Fig. 3). This radiation can be roughly calculated for every day of the year using the following equation:

$$I_0(t) = I_{cs} e(t) \quad [\text{W}/\text{m}^2] \qquad (3)$$

$$e(t) = 1 + 0.033\cos(2\pi n(t)/365) \qquad (4)$$

where $n(t)$ is the progressive number of the day of the year [1].

4 The position of the Sun in the celestial vault

To determine the position of the Sun in the sky at a certain moment of the year and in a certain place, it is necessary to define a few characteristic angles. These angles are [1]:

- the *solar height* or *altitude* α – the angle formed by the direction of the solar rays and their projection on a horizontal plane;
- the *zenithal angle* – the angle formed by the solar rays and the zenith direction; this angle and α are complementary;
- the *solar azimuth* α, which indicates the variance of the solar rays' projection on the horizon's plane as regards the south; by convention, eastward orientations are negative while westward orientations are positive;
- the *hour angle h*, which indicates the angular distance between the Sun and its midday projection along its apparent trajectory on the celestial vault; the time angle is also equal to the angle that the Earth has to rotate to bring the Sun back above the local meridian;
- the *latitude L* – the angle formed by the straight line that connects the place taken into consideration and the Earth's core and its projection on the equator's

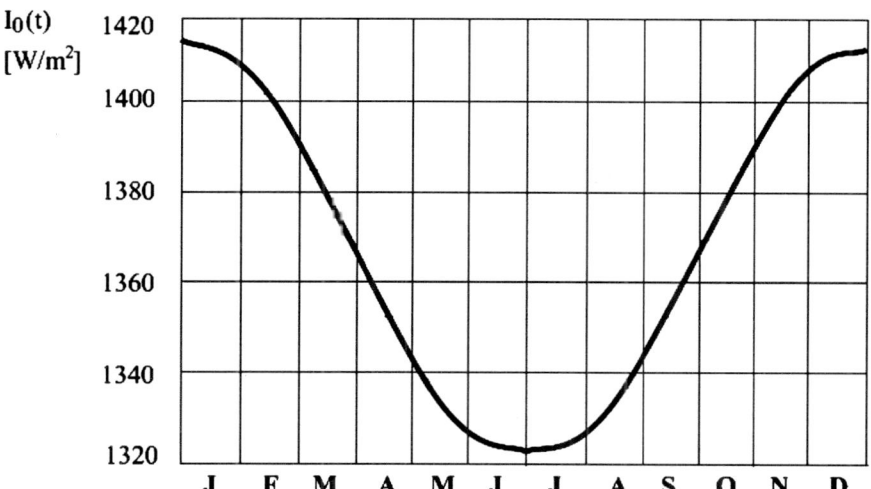

Figure 3: Extraterrestrial radiation $I_0(t)$ during the year.

plane; this angle is positive in the northern hemisphere but negative in the southern hemisphere;
- the *solar declination* δ – the angle formed between the solar ray and the equator's plane measured on the solar midday plane, that is, the meridian plane passing by the Sun; the solar declination is positive when the Sun is above the equatorial plane and negative when it is under the equatorial plane (Fig. 4).

The solar height a and the solar azimuth a define the instant position of the Sun [1]:

$$\operatorname{sen} a = \operatorname{sen} L \operatorname{sen} \delta + \cos L \cos \delta \cos h \tag{5}$$

$$\operatorname{sen} a = \cos \delta \operatorname{sen} h / \cos a \tag{6}$$

Solar declination δ is calculated using Cooper's equation:

$$\delta = 23.45 \operatorname{sen}[360(284 + n)/365] \tag{7}$$

where n stands for the nth day of the year. Declination depends only on the date; therefore, it is the same for all places on the planet.

The hour angle for dawn h_a or sunset h_t can be calculated using eqn (5), avoiding sen α, as [1]:

$$h_a = -h_t = \arccos(-\operatorname{tg} L \operatorname{tg} \delta) \tag{8}$$

8   SOLAR THERMAL AND BIOMASS ENERGY

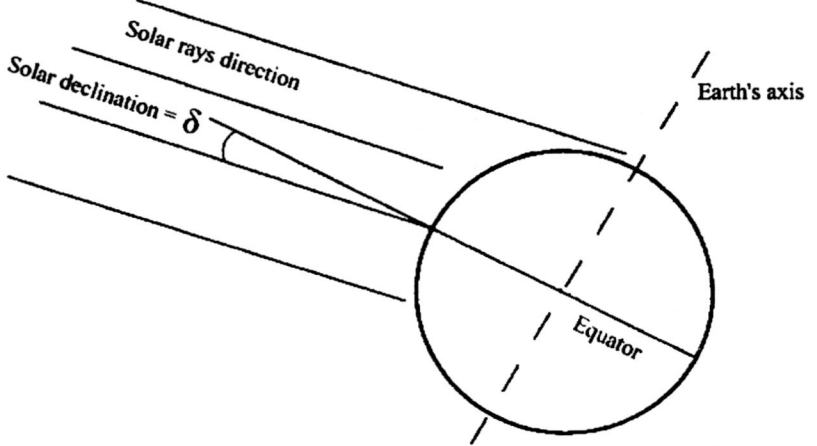

Figure 4: Definition of solar declination.

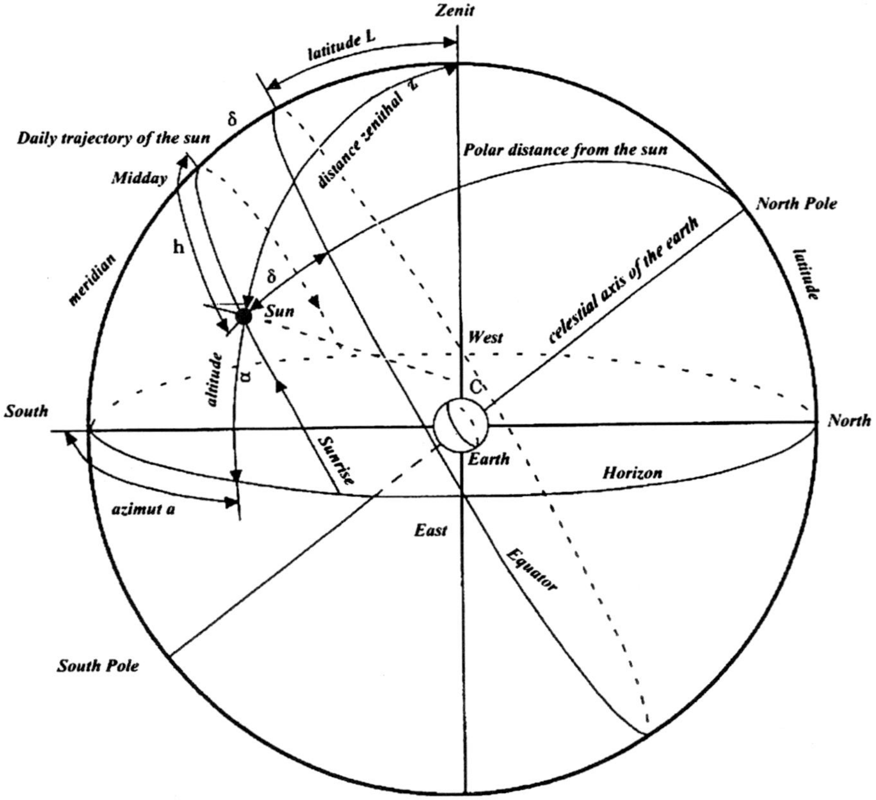

Figure 5: The celestial sphere and solar coordinates for an observer on the Earth at the point C.

5 The solar radiation on the Earth's soil during clear sky days

The solar energy which reaches Earth's surface is much smaller than that which reaches a surface situated outside the atmosphere. This happens because of the phenomenon of diffusion and absorption of solar radiation by components of the atmosphere. The collision with molecules of air, steam and atmospheric dust results in scattered reflection because of which a part of the radiation is sent back to outer space. Absorption, instead, is principally due to ozone (O_3), steam (H_2O) and carbon dioxide (CO_2). O_3 absorbs mainly in the ultraviolet region while H_2O absorbs in the infrared region.

Figure 7 shows the spectral distribution of solar radiation when the Sun is at the zenith.

The part of the solar radiation which reaches the Earth's surface following the direction of the solar rays without being absorbed and reflected is called *directed radiation* (on soil), while the part that reaches the Earth's surface from all directions (because of the scattering) is called *scattered radiation*. *Global radiation* on soil refers to the sum of directed and scattered radiation.

Diffuse radiation can be picked up almost entirely by flat panels since glass is actually transparent to all solar radiation which arrives with an angle of incidence i (i.e. the angle between a solar ray and a normal surface) smaller than the maximum value of reflection (70–80°). On the other hand, concentrators, assuming that they work in conformity with the rules of geometrical optics, have to be oriented towards directed radiation; they do not pick up diffuse radiation.

If we do not consider horizontal surfaces, which are inclined in any manner, besides directed and diffuse radiation, it is necessary to take into consideration a third kind of radiation: the *reflected solar radiation*, the radiation reflected from the soil or from the objects near the given surface; its intensity is influenced by the albedo of those objects. Albedo is the fraction of solar radiation that is received

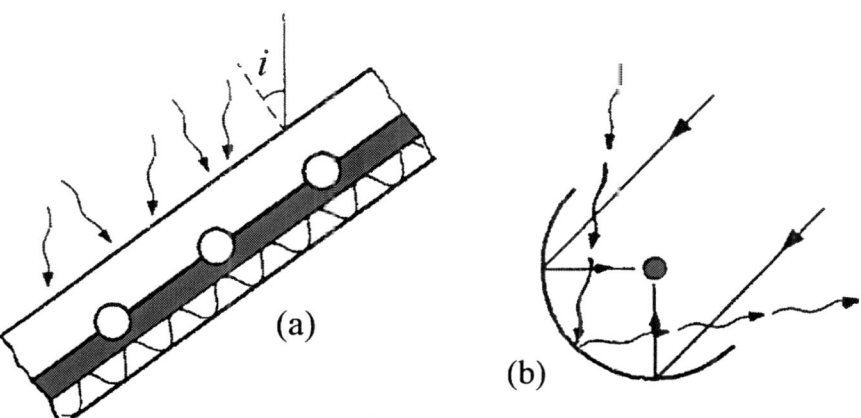

Figure 6: (a) Diffuse radiation picked up by a solar flat panel and (b) directed radiation picked up by a conveniently oriented concentrator.

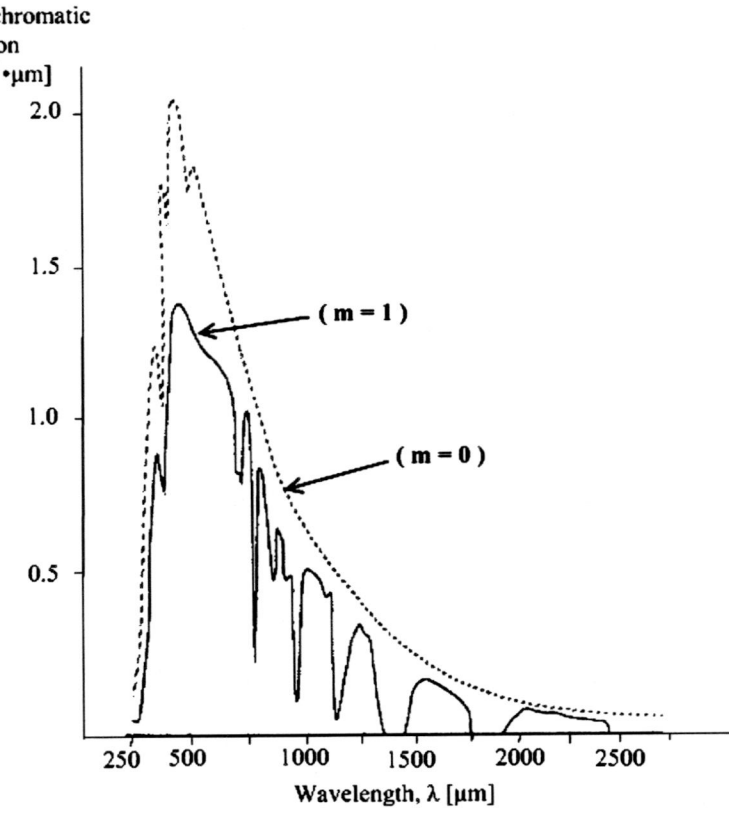

Figure 7: Monochromatic radiation on soil (with the Sun at the zenith, $m = 1$) and outside the atmosphere ($m = 0$).

and suddenly reflected by a surface. Every kind of soil and vegetation has its own value of albedo [1–3].

Albedo can also be defined as a transmission coefficient of the atmosphere, which depends on the wavelength and the route of the solar rays in the atmosphere, besides depending on atmospheric composition, which varies with local weather conditions. In the case of clear sky days, the transmission coefficient of directed radiation, given by the ratio between directed radiation on the soil and extraterrestrial radiation on the orthogonal surface, can be calculated using the following equation:

$$\tau_b = 0.5\{\exp[-0.65m(z,a)] + \exp[-0.95m(z,a)]\} \tag{9}$$

We can assume:

$$m(z,a) = m(0,a)p(z)/p(0) \tag{10}$$

where $p(z)$ and $p(0)$ are the atmospheric pressures at level z and sea level, respectively.

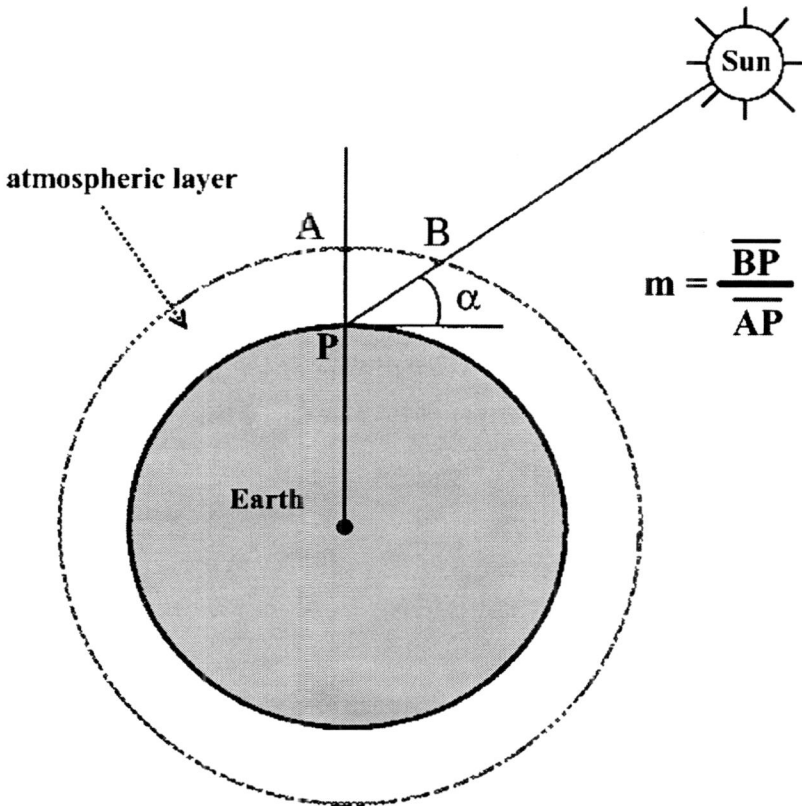

Figure 8: Definition of relative air mass.

The adimensional parameter $m(z,a)$ is the air mass for an altitude z above the sea level. This parameter is defined as the ratio between the effective route length of solar rays and their shortest route length, with the Sun at the zenith; α is the angle formed by solar rays with a horizontal plane (see Fig. 8).

The air mass $m(0,\alpha)$ for the sea level can be calculated using the approximated equation:

$$m(0,a) = 1/\text{sen } a = \text{cosec } a \tag{11}$$

which gives an error percentage of 1% per $a > 15°$, or it can be calculated with the exact formula, taking into consideration the Earth's and the atmosphere's bending:

$$m(0,a) = [1229 + (614\text{sen } a)^2]^{0.5} - 614\text{sen } a \tag{12}$$

The angle a determines the Sun's position in space at any time; the relative air mass m has a certain value; therefore, we calculate τ_b.

Directed radiation is then given by:

$$I_{bn} = I_0 \tau_b \tag{13}$$

Hottel's model is the second way to calculate the radiation on soil during clear sky days. This model estimates direct radiation on clear sky days for a standard atmosphere with 23 km visibility and four kinds of climate.

The transmission coefficient of normal direct radiation (I_{bn}/I_0) is calculated using these relations/equations, which are valid for altitudes lower than 2.5 km:

$$\begin{aligned}\tau_b &= a_0 + a_1 \exp(-k/\text{sen}\alpha) \\ a_0 &= r_0 [0.4237 - 0.00821(6-Z)^2] \\ a_1 &= r_1 [0.5055 + 0.00595(6.5-Z)^2] \\ k &= r_k [0.2711 + 0.01858(2.5-Z)^2]\end{aligned} \tag{14}$$

where Z is the observer altitude expressed in km and r_0, r_1 and r_k are adimensional corrective coefficients.

Table 1: Corrective coefficients of Hottel's correlation.

Kind of weather	r_0	r_1	r_k
Tropical	0.95	0.98	1.02
Summer (average latitude)	0.97	0.99	1.02
Summer (lat. sub-Arctic)	0.99	0.99	1.01
Winter (average latitude)	1.03	1.01	1.00

To achieve global radiation on soil it is also necessary to determine diffuse radiation. Liu and Jordan developed an empirical relation between the coefficient of direct radiation τ_b and that of diffuse radiation τ_d during clear sky days:

$$\tau_d = 0.2710 - 0.2939\tau_b \tag{15}$$

τ_d is the ratio between diffuse radiation on soil over a horizontal plane and extraterrestrial radiation over a horizontal plane (I_0 sen α) [1, 3].

6 Instantaneous direct radiation received on a surface

An inclined surface situated on the terrestrial plane is characterized by two geometrical quantities: *inclination* β, the inclination of the surface compared to the horizontal, and *surface's azimuth* a_w, the angle that the projection on the normal to the surface's horizontal plane has to rotate to superimpose itself on the southern direction. If that rotation is counter clockwise, angle a_w is considered to be positive. The angle between the solar rays and the normal to the surface is called the angle of incidence i.

The direct radiation intercepted by a surface is:

$$G_b = I_{bn} \cos i \tag{16}$$

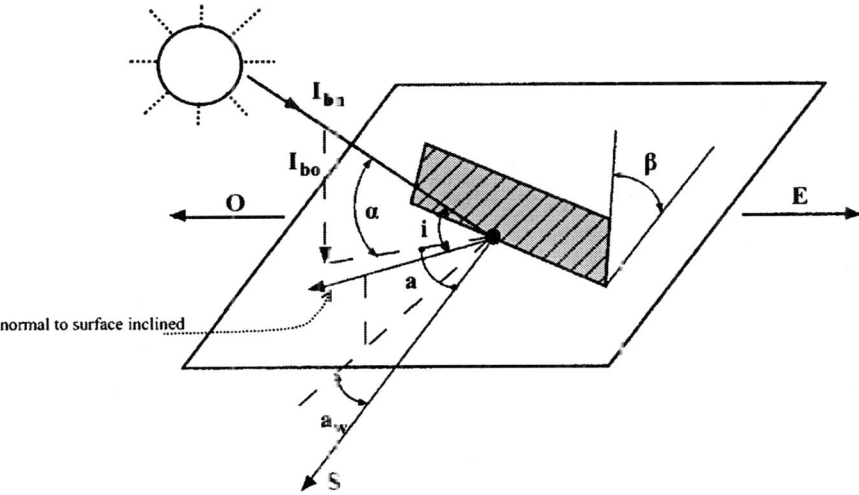

Figure 9: Angles which determine one surface and the Sun's position.

The general expression for cos i:

$$\cos i = \cos(a - a_w)\cos \alpha \operatorname{sen} \beta + \operatorname{sen} \alpha \cos \beta \quad (17)$$

or, as a function of the fundamentals angles L, δ and h:

$$\begin{aligned}\cos i = &\operatorname{sen}\delta(\operatorname{sen} L \cos \beta - \cos L \operatorname{sen} \beta \cos a_w) \\ &+ \cos \delta \cos h(\cos L \cos \beta + \operatorname{sen} L \operatorname{sen} \beta \cos a_w) \\ &+ \cos \delta \operatorname{sen} \beta \operatorname{sen} a_w \operatorname{sen} h \end{aligned} \quad (18)$$

This expression indicates three cases of particular interest:

- *For a horizontal surface* ($\beta = 0°$), we have:

$$\cos i = \operatorname{sen} \delta \operatorname{sen} L + \cos \delta \cos L \cos h = \operatorname{sen} \alpha \quad (19)$$

- *For a vertical surface facing south* ($a_w = 0°$, $\beta = 90°$), we have:

$$\cos i = -\operatorname{sen} \delta \cos L + \cos \delta \operatorname{sen} L \cos h \quad (20)$$

The introduction of the surface's azimuth a_w results in a remarkable complication when compared with the case where $a_w = 0°$. In that case, using geometrical demonstration, it can be easily shown that radiation on an inclined surface of angle β at latitude L is equal to the radiation on latitude $(L - \beta)$, these being surfaces parallels.

Therefore, *for an inclined surface facing south*, we have:

$$\cos i = \operatorname{sen}(L - \beta)\operatorname{sen} \delta + \cos(L - \beta)\cos h \cos \delta \quad (21)$$

Often, it is useful to know when the Sun rises and sets as regards an oriented surface: the surface 'sees' the Sun when the angle of incidence is lower than 90° and the solar altitude is more than 0° at the same time. The Sun rises and sets on the surface connected with the minimum hour angle h_a' (and h_t'), between the absolute value which is calculated by ignoring sen a (hour angle of dawn and sunset on the horizon), and the absolute value is obtained by ignoring cos i (i.e. considering $i = 90°$).

As a rule, in the northern hemisphere for southward oriented surfaces, when it is winter and days are short, it is sufficient to ignore sen a; however, during summer, when the angle of incidence is more than 90° and the Sun has already risen and before it sets, it is enough to ignore cos i. The general rule, valid only if the surface actually sees the Sun, is given by the following equations:

$$\left| h_a' \right| = \min \left| h_a(a = 0°), h_a(i = 90°) \right| \tag{22}$$

$$\left| h_t' \right| = \min \left| ht(a = 0°), h_t(i = 90°) \right| \tag{23}$$

In the simple case of inclined surfaces facing south, we have:

$$\cos i = \cos 90° = 0$$
$$= \text{sen}(L - \beta)\text{sen}\,\delta + \cos(L - \beta)\cos\delta\cos h \tag{24}$$

$$\left| h_a(i = 90°) \right| = \left| h_t(i = 90°) \right| = \arcos\left[-tg\,(L - \beta)\,tg\,\delta \right] \tag{25}$$

$$\left| h_a(a = 0°) \right| = \left| h_t(a = 0°) \right| = \arcos\left[-tg\,L\,tg\,\delta \right] \tag{26}$$

For northward oriented surfaces, there could be both two dawns and two sunsets (in spring and in summer, in the northern hemisphere) and no dawns and no sunsets, that is, absence of direct lightening (in autumn and winter).

When the surface is not oriented southward, it is not possible to get simple closed-form expressions for the hour angles on the surface at dawn and sunset.

Assuming that I_{bo} is the instantaneous direct radiation on a horizontal plane, linked to normal direct radiation by the relation:

$$I_{bn} = I_{bo}/\text{sen}\,a \tag{27}$$

and applying the (16), we get:

$$G_b = I_{bo}\cos i/\text{sen}\,a = I_{bo}R_b \tag{28}$$

where

$$R_b = \cos i/\text{sen}\,a \tag{29}$$

R_b is the inclination factor for direct radiation; remembering that for a horizontal surface it is sufficient to put $\beta = 0°$, expression (28) states that the direct radiation G_b on a surface that is inclined and oriented in any direction is equal to the product of direct radiation I_{bo} on a horizontal plane and the inclination factor [1, 3].

For a southward oriented surface, we have:

$$R_b = \frac{\text{sen}(L-\beta)\text{sen}\delta + \cos(L-\beta)\cos\delta\cos h}{\text{sen } L \text{ sen } \beta + \cos L \cos\delta\cos h} \qquad (30)$$

7 Instantaneous global radiation received on a surface

The instantaneous global power which weighs on an oriented surface is given by the sum of the direct component that is obtained from the eqn (28), the diffuse component, which comes from the celestial vault portion seen from the surface, and the part reflected by the soil and nearby objects towards the same surface.

If the sky's behaviour is assimilated to that of an isotropic spring of diffuse radiation, it is possible to determine the diffuse component which reaches the surface as:

$$G_d = I_{do} R_d \qquad (31)$$

where

$$R_d = \cos^2(\beta/2) = (1+\cos\beta)/2 \qquad (32)$$

where I_{do} is the diffuse radiation on the horizontal plane and R_d is the inclination factor of diffuse radiation.

We can express the radiation (direct and diffuse) reflected by the soil on a certain surface as:

$$(I_{bo} + I_{do})R_r \qquad (33)$$

R_r, the inclination factor of reflected radiation, is equal to:

$$R_r = \rho \text{sen}^2(\beta/2) = \rho(1-\cos\beta)/2 \qquad (34)$$

ρ is the soil's reflection coefficient and it can assume values between 0.2 (grass, concrete) and 0.7 (snow). Therefore, the instantaneous solar power, which is received on a arbitrary oriented surface, in the case of isotropic sky, is equal to [1]:

$$G = I_{bo} R_b + I_{do} R_d + (I_{bo} + I_{do})R_r \qquad (35)$$

8 Calculation of direct energy received on a surface

The direct solar energy received at a certain time interval on an oriented surface situated on the Earth is given by the expression:

$$E_b = \int_{t_0}^{t_0 + \Delta t} I_{bn}(t)\cos i \, dt \qquad (36)$$

where Δt (time interval) may vary (an hour, a day, a month, etc.) and $I_{bn}(t)$ stands for the normal direct radiation.

As a rule, it is not possible to use this equation to calculate E_b since $I_{bn}(t)$ depends on local atmospheric conditions which cannot be known in advance. It is possible, instead, to calculate that quantity per surface on soil during clear sky days using one of the models described in par. 5.

It is often useful to calculate using eqn (36) the radiation received during the day on a horizontal surface placed outside the atmosphere, which is equal to:

$$H_{ex} = \int_{h_a}^{h_t} I_{cs} e(t) \operatorname{sen} a \, dt \tag{37}$$

where h_a and h_t are, respectively, the hour angles at dawn and sunset. Through a few passages we get:

$$H_{ex} = 24/\pi I_{cs}[1+0.0033\cos(2\pi n/365)] \\ \cdot (\cos L \cos \delta \operatorname{sen} h_a + h_a \operatorname{sen} L \operatorname{sen}\delta) \tag{38}$$

where h_a, in the last term, should be expressed in radians. The daily extraterrestrial radiation calculated using eqn (38) is expressed in watt-hour per square metres (W-h/m^2) [1].

9 The true solar time

The hour angles which correspond to the different positions of the Sun in the sky concern the true solar time and not the conventional time measured by a clock. However, true solar time is determined by the Sun's position in the sky: for example, when the Sun is on the meridian of a certain place, the hour angle is nil because of the solar midday; instead, the clock will indicate a different time.

To convert clock time to true solar time, first, it is necessary to correct the longitude difference between the local meridian and the standard time zone meridian, considering that any grade of longitude difference corresponds to a four-minute correction; second, because the angle speed of the Earth is not constant during the year, it varies positively or negatively as regards the average conventional value of 360/24 grad/h, we have to correct it using the equation of time (ET).

We use the relation [1]:

True solar time = Conventional time + 4′(Longitude of the local meridian − Longitude of the standard time zone meridian) + ET

If we apply legal time, the conventional hour must not be the legal one, but the one for the standard time zone meridian. Longitudes are considered to be positive if they are eastward from Greenwich. As regards Italy, the datum meridian is at 15° east of Greenwich and it passes by the volcano Etna. Figure 10 shows the ET trends during the year.

Some values of ET are shown in Table 2.

The Solar Radiation 17

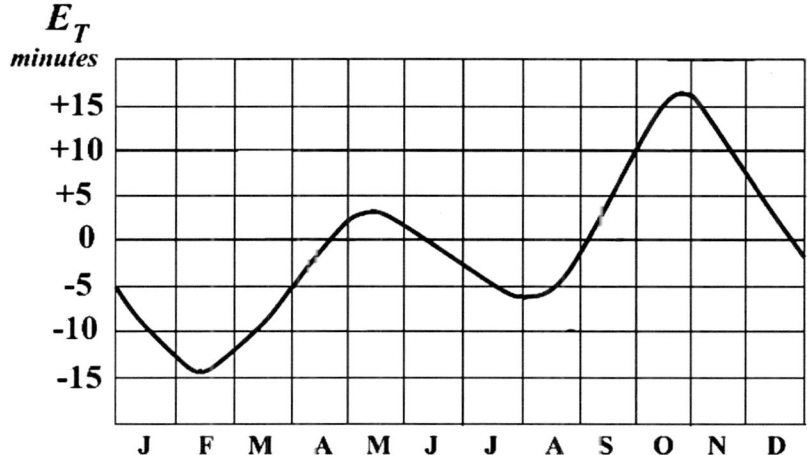

Figure 10: Trends for equation of time during the year.

Table 2: Values of the equation of time during the year.

Date		ET		Date		ET	
Months	Days	Min.	Sec.	Months	Days	Min.	Sec.
Jan.	1	−3	14	Jul.	1	−3	31
	13	−8	27		13	−5	30
	29	−13	5		29	−6	23
Feb.	1	−13	34	Aug.	1	−6	17
	13	−14	20		13	−4	57
	29	−13	19		29	−1	10
Mar.	1	−12	38	Sep.	1	0	15
	13	−9	49		13	3	45
	29	−5	7		29	9	22
Apr.	1	−4	12	Oct.	1	10	1
	13	0	47		13	13	30
	29	2	33		29	16	10
May	1	2	50	Nov.	1	16	21
	13	3	44		13	15	47
	29	2	51		29	11	59
Jun.	1	2	27	Dec.	1	11	16
	13	0	18		13	6	12
	29	−3	7		29	−1	39

10 The diagram of solar trajectories

Using a polar diagram, it is possible to visualize solar trajectories during a year at a certain place. This diagram, which is a projection of solar trajectories on a horizontal plane, can be obtained by the plotting the values of solar height and azimuth on a graph. These values are calculated using the eqns (5) and (6), for a certain place and as a function of the true solar time, as shown in Fig. 11.

Using this diagram, it is also possible to determine graphically the periods of time during which a surface point remains in shadow because of the obstacles which intercept solar rays. When the distance of the obstruction is large compared to the receiver's dimensions (a solar collector, a window, etc.) it is right to consider the receiver as a punctiform one, since the shadow tends to move fast on the receiver so that it is completely in shadow or completely illuminated. To determine when the obstacle intercepts solar rays, in the diagram of solar trajectories, we have to represent the angle from the obstacle as seen from the considered

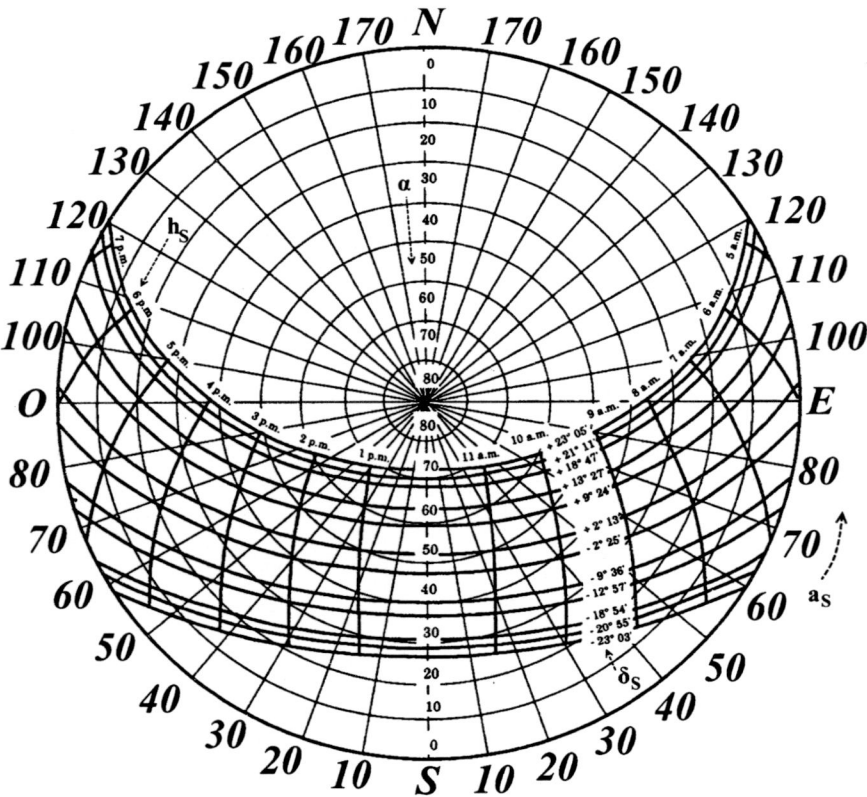

Figure 11: Diagram of solar trajectories.

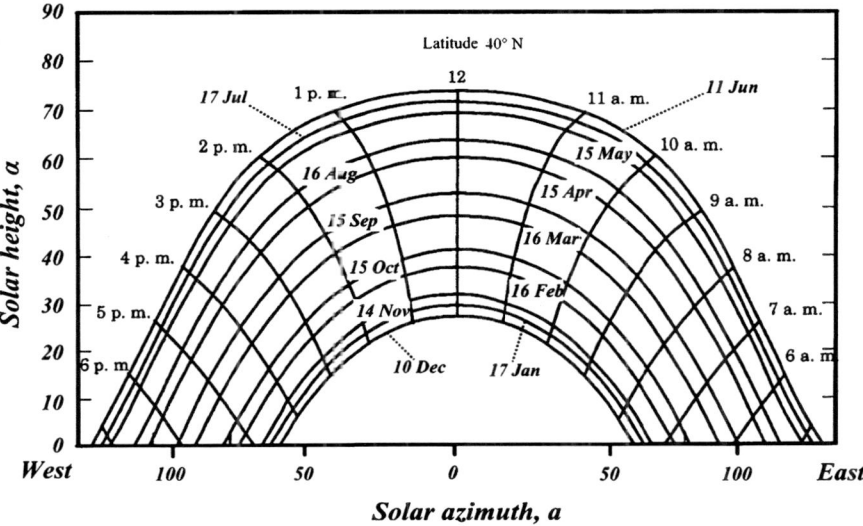

Figure 12: Diagram of the Sun's position.

point, plotting on it (the diagram) the azimuth and the angle height of the obstacle's contour points.

As an alternative to the diagram of solar trajectories we can use a Cartesian diagram of the Sun's position in which the azimuth is plotted along the horizontal axis while the altitude is plotted vertically. The Sun's position can be read by simply reading the two axes. An example of this diagram is given in Fig.12. Of course, we may use this diagram to calculate the shadows [1].

11 The monthly average solar radiation on inclined surfaces

Starting from the experimental values obtained on horizontal surfaces, Liu and Jordan have introduced a widely used method to calculate the monthly average solar radiation on inclined surfaces. This method is based on the division of radiation between its direct and diffuse components. Liu and Jordan discovered that the ratio between the monthly average diffuse radiation D and the global radiation H received on a horizontal surface can be correlated to a parameter called monthly clearness index K. This index is obtained by dividing the monthly average terrestrial radiation for every day by the monthly average extraterrestrial radiation for every day both received on a horizontal plane.

$$K = H/H_{ex} \tag{39}$$

$$D/H = 1.39 - 4.027K + 5.531K^2 - 3.108K^3 \tag{40}$$

To compute H_{ex} the solar constant value used is 1394 W/m² (instead of the more recent value of 1367 W/m²). For this reason, K values should be based on that value. As for the calculation of H_{ex}, it has been suggested that eqn (38) be applied to a specific day of each month. That day must be chosen to get an extraterrestrial radiation H_{ex} on a horizontal surface, which is equal to the monthly average extraterrestrial radiation H_{ex}.

If $B (= H - D)$ is the monthly average direct component received on a horizontal surface for each day and E is the monthly average global radiation on an arbitrary oriented surface, we have:

$$E = R_b\,B + R_d\,D + R_r\,(B+D) \tag{41}$$

$$R_b = \frac{\cos(L-\beta)\cos\delta\,\mathrm{sen}\,h_a' + h_a'\mathrm{sen}(L-\beta)\mathrm{sen}\delta}{\cos L \cos\delta\,\mathrm{sen}\,h_a + h_a \mathrm{sen}\,L\,\mathrm{sen}\delta} \tag{42}$$

where h_a' is the hour angle calculated using eqn (22); h_a and h_a' are expressed in radians and R_b is the monthly average factor of inclination by direct radiation [1].

12 Daily radiation on an inclined surface

Liu and Jordan extended their model of division, which is valid for monthly average radiation values, to the valuation of daily global radiation received on an inclined surface. If K is the daily clearness index, which is equal to H/H_{ex} (i.e. the ratio between daily radiation and extraterrestrial radiation received on a horizontal surface on a specific day), the ratio between diffuse radiation and global radiation received on a horizontal surface on a certain day has to be calculated as follows:

$$\begin{aligned}D/H &= 1.0045 + 0.04349K - 3.5227K^2 + 2.6313K^3 \quad \text{se } K \leq 0.75 \\ D/H &= 0.166 \quad \text{se } K > 0.75\end{aligned} \tag{43}$$

The direct component is calculated by the difference:

$$B = H - D \tag{44}$$

The daily global radiation on an arbitrary inclined surface can be calculated using the following expression:

$$E = R_b B + R_d D + R_r (B+D) \tag{45}$$

where R_b, R_d and R_r are the inclination coefficients of direct, diffuse and reflected radiation determined, respectively, by eqns (29), (32) and (34), but for the chosen day [1].

13 Hourly solar radiation on inclined surfaces

The hourly values of global radiation received on a horizontal surface H_h can be divided into diffuse components D_h and direct components B_h by Liu and Jordan's method using the following expressions:

$$D_h/H_h = 1 - 0.09k \qquad \text{if } k \leq 0.22$$
$$D_h/H_h = 0.9511 - 0.1604k + 4.388k^2$$
$$\qquad - 16.638k^3 + 12.336k^4 \qquad \text{if } 0.22 < k < 0.8 \qquad (46)$$
$$D_h/H_h = 0.165 \qquad \text{if } k \geq 0.8$$

where k is the hourly clearness index defined by the ratio between the hourly global energy H_h received on a horizontal plane and the hourly energy received on a horizontal plane $H_{h,ex}$ situated outside the atmosphere.

$$k = H_h/H_{h,ex} \qquad (47)$$

$H_{h,ex}$ can be calculated using the equation:

$$H_{h,ex} = I_{cs}[1 + 0.033\cos(2\pi n/365)]$$
$$\cdot (\cos L \cos \delta \cos h + \sen L \sen \delta) \qquad (48)$$

The hour angle h is calculated at the centre of the considered hour or using the exact equation:

$$H_{h,ex} = 12/\pi I_{cs}[1 + 0.033\cos(2\pi n/365)]$$
$$\cdot \{\cos L \cos \delta (\sen h_1 - \sen h_2) + (h_1 - h_2)\sen L \sen \delta\} \qquad (49)$$

h_1 and h_2 are the hour angles at the extremities of the said hour; in eqns (48) and (49) $H_{h,ex}$ is expressed in W-h/m^2.

Once we have obtained the value of the hourly global radiation H_h received on a horizontal plane, we can calculate the hourly diffuse radiation D_h. The hourly direct radiation on a horizontal plane is calculated by difference:

$$B_h = H_h - D_h \qquad (50)$$

The hourly global radiation received on an inclined surface turns out to be:

$$E_h = R_b B_h + R_d D_h + R_r (B_h + D_h) \qquad (51)$$

In this case, R_b is calculated at the centre of the said hour [1].

14 The local radiation data retrieval

As regards Italy, there are medium-height solar radiation regimes with a big variation between northern and southern regions. Keeping in mind that the parameters required to determine univocally the position of an intercepting surface are the surface's inclination and its azimuth orientation, we now list the principal sources for the retrieval of radiation data.

'La radiazione globale al suolo in Italia' [10], a paper edited by ENEA, is without any doubt the bibliographic source which supplies, on a national level, the most detailed information about the average global radiation (i.e. the one which includes direct, diffuse and reflected components) received on a square metre of a horizontal surface per month and year. The same kind of information, which refers to a smaller number of Italian places, is also provided by UNI ISO 10349 international norms.

'L'atlante europeo della Radiazione Solare' [11] is without any doubt one of the most authoritative sources for the valuation of solar radiation received in a certain period of time on a surface which is exposed in any manner. This atlas gathers all the data supplied by national metrological offices. These data, gathered in maps and tables, are the result of a 10-year study. The atlas is divided into two volumes: the first takes into consideration the horizontal surfaces and the second the inclined surfaces. The first volume reports, for every Italian place we consider, the values of the daily average radiation expressed in W-h/m^2 or in kW-h/m^2. Every place is characterized by its latitude, longitude and height above sea level.

As regards the design of solar panels, the results of the second volume appear to be much more interesting than the first. As a matter of fact, solar panels are usually arranged with a certain inclination on a horizontal plane, and the second volume reports the values of the daily average radiation (global and diffuse) per month and year for different positions of the intercepting surface. However, in the European Atlas of Solar Radiation, only the principal cities of each country are mapped.

To obtain the values of radiation received on variously oriented and inclined surfaces, there are a few algorithms (which can be easily found in the currently available design software) among which the most well known and used is that of Liu and Jordan which is discussed in par. 11, 12 and 13. The steps to get a correct extrapolation of the data for a surface which is positioned in any manner, starting from the values for a horizontal surface, are outlined in the UNI 8477 norms (first part) [4, 5].

15 Variation in the energy which can be intercepted by the position of the surfaces

Many factors affect the positioning of the solar system's intercepting surfaces. Among them, the most important is the study of the place and users' requirements; actually, it is important to examine carefully the consumptions trend during the year. Another factor that has a strong impact on the positioning of the intercepting surfaces is the shading phenomenon at the installation site. To determine accurately the shades which can appear on a certain surface, we can use the solar trajectories diagram or the Sun's position diagram (par. 10), which provides precise information on the Sun's position in the celestial vault during the day and the year at a certain place.

It is clear that the best orientation for an intercepting surface is the one that is orthogonal to the solar rays. The fixed intercepting surfaces (i.e. the ones that do not have automatic Sun chasing devices) meet that orthogonality condition once a day. Hose surfaces are normally installed southward to maximize the energy received during the day. However, this is not a strict norm, especially where the roof is not north–south oriented. Panels which are eastward or south-eastward oriented favour the morning running while the westward or south-westward oriented panels favour the afternoon running.

Choosing the best inclination is not easy and immediate; generally, it is chosen such that it is equal to the latitude L decreased by about 10° to maximize the energy collected during the year (e.g. in Rome since the latitude is 42°, the best inclination will be 30°).

If users require the system to work especially in winter months, this value will not be satisfactory. During winter, the apparent trajectory of the Sun in the celestial vault is on average low so that the average inclination of solar radiation reaches the minimum yearly values. A panel inclination higher than the mentioned 30° (e.g. 60° for a hotel located in a skiing resort) will be necessary to favour the intercepting surfaces' exposure to direct radiation.

On the contrary, in summer (e.g. for an open air swimming pool) users can maximize the service with an inclination of about 10°.

The last factor which contributes to the correct positioning of surfaces is the economic result of the investment: the right dimensions and the correct realization of the system minimize the need for the active surface and therefore the number of collectors to be bought and the overall cost of the operation. Eventually, it is necessary to point out that small positioning variations compared to the best panel positioning can lead to negligible loss of energy received [2, 5].

CHAPTER 2

Solar energy utilization

1 Introduction

There are two principal ways of solar energy exploitation:

- heat production (for use in the domestic, civil and production fields; in this case, we talk about *thermal solar energy*);
- electricity production by the direct conversion of energy (*photovoltaic solar energy*).

Thermal solar technologies are divided into low, medium and high temperature ones.

The low temperature technology includes systems which thanks to appropriate devices (solar collectors, see par. 2.2.2.1), are able to heat fluids at temperatures less than 100°C. These systems are generally installed to produce sanitary hot water (for domestic use, collective users, sport centres, etc.), to produce domestic heating and, in general, other room heating, to heat water in swimming pools, to produce heat at a low temperature for industrial utilization (usually to warm the water used to swill machines or to preserve different kinds of fluids at a certain temperature inside tanks, etc.).

The medium temperature technology includes systems, which allow reaching temperatures of more than 100°C and less than 250°C. Currently, medium-temperature solar thermal power systems are not widespread; among their applications, the most common application is the one represented by simple devices which use solar radiation to cook food (the so-called solar ovens, see par. 2.3).

The high temperature technology includes systems which, thanks to appropriate devices that are able to concentrate solar radiation to a thermal receiver (in this case we talk about concentrating solar power (CSP) technology, see par. 2.4), allow heating a fluid at temperatures more than 250°C.

Concentrated solar technology has its application in electricity production (in this case, we talk about 'solar thermodynamics' where the heat at a high temperature is exploited in thermodynamic cycles for electricity production) and in the fulfilment of chemical processes at high temperatures, such as production of hydrogen. [2, 12–15].

2 Low-temperature solar thermal technology

Based on thermodynamic considerations, the use of electrical energy to produce hot water is not recommended, since a prized kind of energy is used and also because the global efficiency of the water heating process is lower than the production of many other direct water heating processes. In fact, heat is a kind of energy which we inevitably find in every real process as consequence of the irreversibility of this process. So it does not make sense to degrade completely a noble form of energy to obtain heat, without getting the mechanical work which can be obtained from that energy.

An alternative way to produce hot water involves the exploitation of solar energy, which represents a form of clean and inexhaustible energy, by low-temperature thermal solar technologies [16].

These technologies include systems using a solar collector to heat a fluid or the air. The aim of these low-temperature thermal solar systems is to intercept and transfer solar energy to produce hot water or heat buildings. By low temperature we mean the heating of fluids at a temperature of less than 100°C (it rarely reaches 120°C) [1, 2, 5, 13, 17].

The detailed description of low-temperature thermal solar systems can be found in par. 2.2.2.

2.1 The advantages

The rational justification for a low-temperature thermal system derives from economic and environmental reasons. The reduction of environmental pollution and the saving of energy which can be obtained using solar energy represent a solid advantage for the community.

The Sun can normally give us 80–95% of the hot water that we use daily to wash our hands, take a shower, do the dishes or also wash clothes, if we manage to connect the solar system to the dishwasher's or washing machine's hot water entrance. If the products are of good quality and the system is well sized, the outcomes are very good even in northern Italy. While in southern Italy the outcomes are excellent in winter also, in northern Italy we can achieve good results only if one more panel is installed or if a capacious tank is provided to make up for the lack of sun during cloudy days, since a good tank keeps water at a constant temperature for a few days. It is not easy to know the exact saving that could be obtained from a sanitary water solar system, as it depends a lot on the habits of families. So it depends on how much water is used for personal hygiene or for dishwashing and also on the kind of central-heating boiler or water heater the family possesses. Using an approximate calculation, which is sufficiently reliable, we can indicate a yearly methane saving of 100–180 m^3 per person, with a yearly lack of carbon dioxide emission (the main cause of greenhouse effect) of 230–400 kg per person. To these quantities we have to add the wastes generated by the accumulation of the central-heating boiler and water heater, for example, with a simple pilot flame or with intermittent running of the water heater to keep the temperature of a huge quantity of water constant during the more unthinkable hours of the day and night. These methane wastes will be 150–200 m^3 per year and they have to be added to the real savings calculated on the basis of the number of persons in a family. The methane cost per m^3 or the cost for the fuel used by our own central-heating boiler has to be taken into consideration to calculate correctly the saving in conformity with taxes, the additional tax when exceding the permitted use of energy, the IVA, the inflation, the rise which will take place in the next few years, etc. Since solar systems are capable of producing more than 90% of the hot water requirements between April and October, the central-heating boiler or the water heater could be turned off during this period, resulting in a significant reduction of wastes.

It is important to determine the regeneration time of the investment, which may be important to justify a solar system installation. The number of years taken to regain the investment can be obtained by dividing the cost paid by the yearly maximum money saving obtained by the use if sanitary hot water produced by solar energy. On average, if we talk about a single family system, the hot water cost alone will be about 1500–2600 €, amortizable within 3–5 years, while the system's useful technical life span is about 20 years and the maintenance costs are 1% of the system's original cost.

A traditional water heater (electrical or methane one) never regains its cost because there is always a bill cost, whereas the more a solar energy system is used, the more it is convenient.

Only when the use of water is reduced, it is more convenient to have a small electrical or gas water heater [16, 18].

2.2 Low-temperature solar thermal system

At our latitudes, low-temperature solar thermal systems are more common than the medium and high temperature ones (especially among private users, who want to save money on their energy bills). They are usually used [1–5, 17]:

- to heat sanitary water for domestic, hotel and hospital use;
- to heat water for showers (bathing establishments, camping, etc.);
- to heat rooms;
- to heat water used in processes at a low temperature;
- to dry foodstuffs;
- to refresh rooms (although it is still too expensive).

A solar thermal system always works in the same way, although there may be slight changes according to its application and use. First, there are a few solar collectors which absorb large quantities of sunlight and then convert it into heat.

The collectors are crossed by a fluid (thermal vector fluid) which removes the absorbed heat; this fluid, crossing the so-called solar circuit, arrives at an accumulator which stores the large quantities of thermal energy to be used in the future when there will be a real need [17].

We will now analyse the low-temperature solar thermal system starting with its most important element: the collector.

2.2.1 The collector

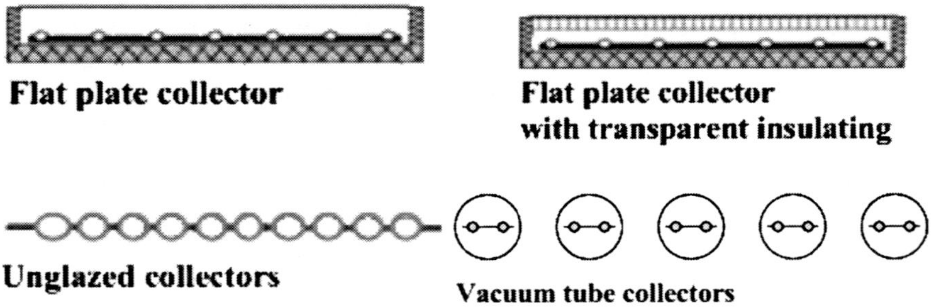

Flat plate collector

Flat plate collector with transparent insulating

Unglazed collectors

Vacuum tube collectors

Figure 1: Sections of different kinds of collectors.

Currently, there are four principal types of solar collectors which have been studied to get the best ratio between costs and benefits according to the different conditions for their application and their possible uses [1–6, 12, 17]:

- flat plate collectors (very common, medium cost, versatile);
- vacuum tube collectors or evacuated tube collectors (high efficiency, more expensive, but useful during any time of the year);
- unglazed collectors or pool collectors (only for use in the hot season, generally for swimming pools or bathing establishments, very economical);
- integrated storage collectors (useful in mild climate zones, they decrease the cost of the solar system).

2.2.1.1 Flat plate collectors Let us analyse the working principle of a generic flat plate solar collector used to heat a fluid. Every device included in this category aims to convert the maximum part of the electromagnetic energy received with the solar radiation into thermal energy, which is available to the users. To serve this purpose, we exploit and strengthen the capacity of certain materials, for example, metals such as copper and alloys such as steel, to warm up fast when exposed to solar radiation and to release the stored heat very easily. The most important element of solar panels is the absorber plate, which has the above-mentioned characteristics; this plate is crossed by tubes through which the fluid that has to be warmed up flows.

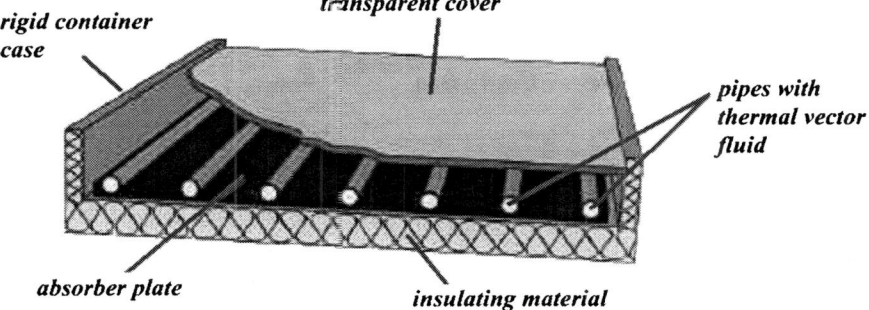

Figure 2: Structure of a flat plate collector.

Figure 3: Flat plate collector.

Figure 4: An installed flat plate collector.

All the mechanisms for exchanging heat from the plate-tube elements to elements which are not the fluid have to be minimized or reconverted to transfer the greatest quantity of the solar radiation received to the fluid.

For this reason, the posterior part of the plate, the part which is not exposed, and its side parts are lined with insulating materials, and the temperature inside the collector is also kept at its highest level, thanks to one or more covering transparent plates [5].

Let us now see the collector's working in detail.

The solar panels in Fig. 7 have a structure that comprises a rigid container case insulated on the inside. A transparent cover reduces energy losses to the outer side of the collector and favours penetration of the received radiation which is intercepted by a black metal plate situated below (intercepting or absorber plate) [2].

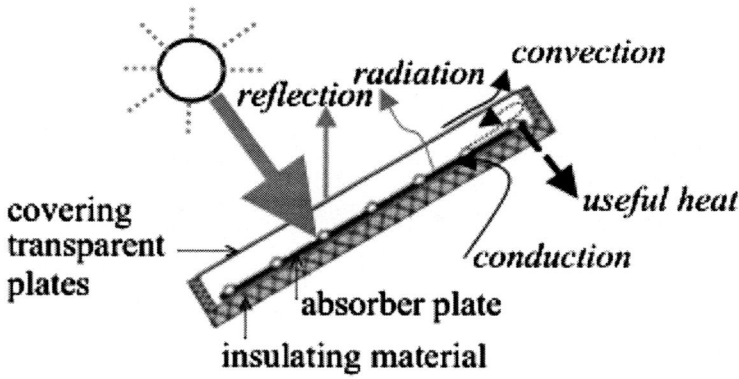

Figure 5: Flat plate collector.

Figure 6: Flat plane plan collectors.

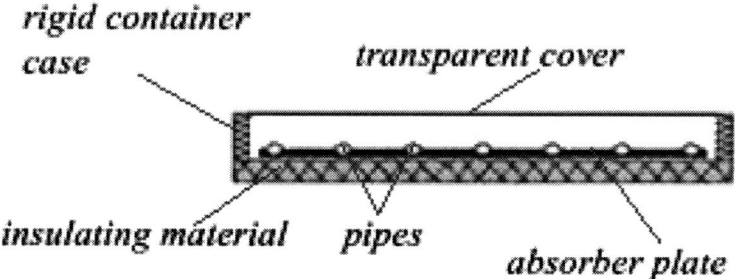

Figure 7: Section of a flat plate collector.

The *thermal vector fluid* flows inside the pipes, which are in contact with the metal surface, and removes the absorbed heat. Sometimes, the pipes through which the fluid flows are welded on the plate; however, most of the time, canalizations are made as shown in Fig. 8, that is, using the *roll bond* method, which comes from refrigerating technology. (This method consists of welding two plates by hot-rolling; but before this, the worm-pipe design – to allow the flow of the thermal vector fluid – has to be printed on one of these plates by the silk-screen process.) Canalizations are normally able to resist pressures of 6–7 bar, although some collectors can guarantee a resistance of pressures even up to 10 bar.

Figure 8: Pipes through which the thermal vector fluid flows, obtained using the roll bond technology between two welded metal plates.

Usually, the most commonly used thermal vector fluid is a mixture of water and propylenic glycol (non-toxic and a good antifreeze), but depending on the application either water alone (which poses two major problems, namely lime scale accumulation and a very high freezing point) or simple saline solutions can also be used. The hydraulic circuit of the panel is depicted in Fig. 9.

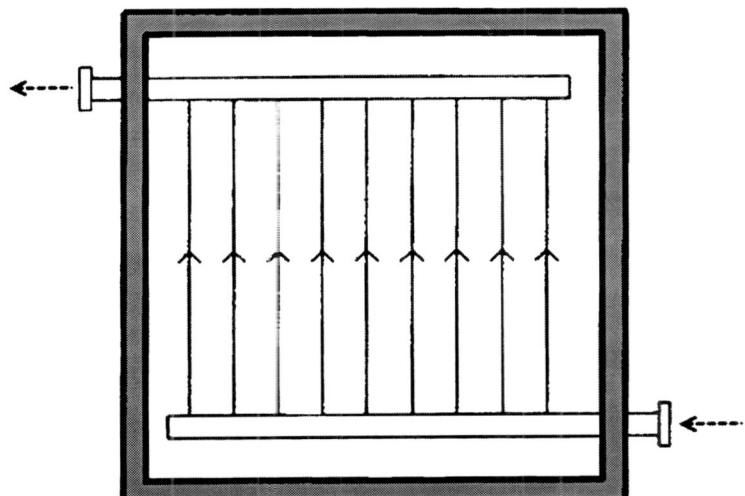

Figure 9: Hydraulic circuit of a solar panel.

The most important requirements of a thermal vector fluid are:

- high density and high specific heat (to use pipes of small dimensions);
- it should not corrode the walls of the circuit;
- chemical inertia and stability at a temperature of less than 100°C;
- restrained hardness to limit lime scale accumulation;
- low freezing point;

- low viscosity;
- it must not be toxic (in the case of sanitary hot water supply). As regards room heating, since the requirement of non-toxicity is not essential, a mixture of water and ethylic glycol (which has a thermal capacity higher than that of propylic glycol) can also be used.

If we are going to use water mixtures, it is important to prevent the freezing of these mixtures using antifreeze solutions. As a matter of fact, when the days or nights are very cold and there is lack of solar radiation, the liquid may freeze and in the process it expands and may break the collectors or the solar circuit [2, 5, 8].

As we will see in par. 2.2.1.6, there are some collectors which use air as the thermal vector fluid instead of a liquid.

The *intercepting plate* (or absorber plate) should be made of a metal that has a low thermal resistance. For this reason, the most commonly used plates are made of copper (the best ones), aluminium (the next best ones) and steel. The absorber plate is covered on the outside by a dark finish coating (as we will see later, because of the kind of finish there can be selective or non-selective surface panels). When solar radiation hits the absorber plate, the radiation is almost completely absorbed, while only a small part is reflected. The absorbed radiation produces heat which is transferred by the sheet to the copper pipe through which the thermal vector fluid flows; finally, this fluid absorbs the heat. The quantity of the reflected radiation has to be restricted as much as possible because it has the same characteristics as the received radiation and it is unsuitable that a large part of the radiation is returned to the atmosphere.

Figure 10: 'Front–back' view of an absorber plate.

Steel plates, besides having a low thermal conductivity, have a high thermal capacity and hence they are less efficient in exploiting the thermal transitories which are connected to the passing of the clouds. If we use aluminium plates, we have to insert dielectric joints inside the hydraulic circuit, which generally has copper elements, to avoid corrosion due to the creation of copper–aluminium piles [2, 8].

There are two kinds of plate panels [2, 5, 8, 17, 21]:

- Those with a *non-selective surface*: The absorber plate surface is treated with dark mat paints. These kinds of paints reduce the losses due to reflection and increase the plate's ability to absorb the wavelengths of solar radiation. This panel is recommended for holiday houses, as hot water is used only in summer and it takes care of the supply of sanitary hot water. If there is a good quality tank, then 100% supply of sanitary hot water can be easily achieved.
- Those with a *selective surface*: The heat absorber is potentiated by a surface which allows the panels to combine the non-selective surface characteristics (reduction of reflected radiation losses and high ability to absorb wavelengths of solar radiation) with a low emissivity for wavelengths that characterize the infrared radiation, which is characteristic of a body at a temperature of nearly 100°C. As a matter of fact, although the plate can count on the covering opacity as regards infrared radiation (which remains inside the collector, see Fig. 11), it is important that the absorber plate emits the smallest quantity of energy to the atmosphere by radiation, since a part of the absorbed energy is dispersed in any case (e.g. one of the flows in Fig. 11 represents the quantity of energy absorbed by the covering and then emitted outside). This kind of panel is much more efficient and expensive, but it can be used during all 12 months of the year. Table 1 lists a few examples of selective surfaces.

Table 1: Examples of selective surfaces.

Coating	Substrate	Solar absorption (a)	Infrared emissivity (ε)
Black nickel on nickel	Steel	0.95	0.07
Black chrome on nickel	Steel	0.95	0.09
Black chromium	Copper	0.95	0.14
Black chromium	Steel	0.91	0.07
Iron oxide	Steel	0.85	0.08
Manganese oxide	Aluminium	0.70	0.08
TiNOX	Copper	0.95	0.04

Non-selective surface panels have a higher infrared emissivity (on average $\varepsilon = 0.85$) than selective surface panels, resulting in higher leakage of useful energy. The best material used to construct an absorber plate is a thin copper sheet lined with a TiNOX selective material (a titanium and quartz covering released in the market in 1995; it does not contain either chrome or nickel). As one can observe from Table 1, besides a high level of solar absorption, TiNOX has the lowest emissivity per wavelength of infrared radiation.

The *container case* should provide compactness and mechanical solidity to the collector and should also protect the inner elements from dirt and atmospheric agents. The container case should be perfectly waterproof to prevent humidity from entering te collector; otherwise, the humidity that enters evaporates as soon as it comes into contact with the hot plate and if the outside temperature is low, it condenses against the inside face of the glazing reducing its transparency. Moreover, humidity can raise the thermal conductivity of fibrous materials (such as wool, polyurethane, polyester wool or stone wool) which are used for internal insulation of the panels. The container case is made of stainless steel (generally, zinc-plated or a pre-treated one), anodized aluminium or, more rarely, fibreglass [2, 5].

To increase the penetration of the radiation received to its maximum and to restrict the energy losses to the atmosphere, the *collector's covering* has to be transparent to the wavelengths of the solar radiation (on average 0.2–0.5 µm) and, at the same time, it has to be opaque to the infrared radiation which comes from the pipes and plate taken together while their temperatures increase gradually (wavelengths higher than 4 µm). Glass meets these requirements best especially if it is treated to get more resistance and transparency (generally, two sheets of tempered, prismatic and antireflection glass are used). Nevertheless, because of the fragility of glass and its weight, sheets of plastic materials (such as polycarbonate) are preferred to glass. Glass provides an additional level of security because if it breaks, it breaks into very small but not sharp parts, thereby reducing the risks of accidents.

The transparent covering is the Achilles' heels as regards the thermal losses to the atmosphere; it is the only surface that cannot be insulated in a proper way [5].

The panel is insulated to avoid conductive losses towards its back and sides and thanks to the insulation they are negligible. Moreover, to create microscopic air spaces (which are good barriers to heat transmission), the different utilizable materials used (polyurethane, polyester wool, fibreglass or stone wool) are always characterized by a porous or alveolar structure. To fight humidity, insulating materials are often covered with a very thin aluminium sheet, which, at the same time, reflects towards the absorber plate, the energy that it receives from the same plate by radiation [5].

Figure 11 shows the thermal flows inside a collector.

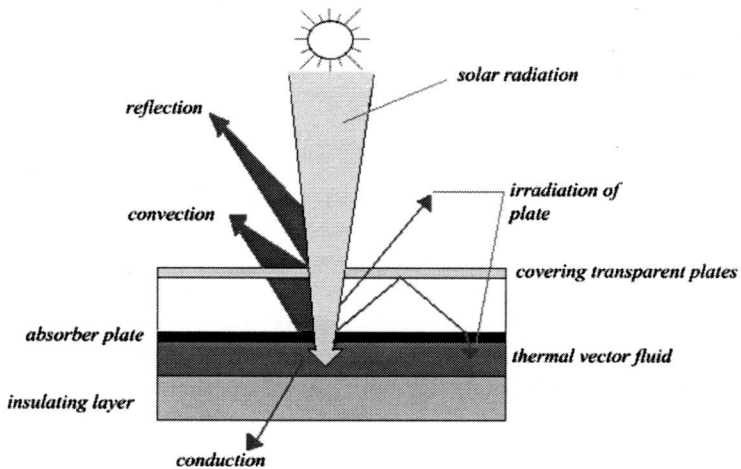

Figure 11: Thermal flows inside a collector.

Compared to conductive losses (the choice of a good insulating material can effectively restrict the losses towards the back and the sides of collector), the losses due to convective motions in the air space between the absorber plate and the transparent covering are difficult to contain. This causes damage to the collector's performance, especially in places where the temperatures are low during most part of the year. When air comes into contact with the plate, it warms up quickly and tends to move up transferring a substantial part of its heat to the cover, which, as it is made of non-thermal insulated materials, then allows the heat to follow its natural course towards the environment which has a lower temperature (i.e. the outside).

To restrict this heat loss, we can use two glazings: the still air space (or better the insulating gas space) between the two glazings forms an efficient barrier against the escape of heat. However, in panels with double glazing, the flow of received radiation decreases because the limit angle i_l, the angle above which glass becomes reflecting, is small.

Another solution would be to use a transparent surface made of alveolar polycarbonate, but, although it is more light, handy and cheap, its optical properties tend to deteriorate more quickly than that of glass [2, 5].

Collector's efficiency The collector's efficiency is determined by the ratio of the energy acquired from the thermal vector fluid and the energy received on the collector's surface at a certain time unit:

$$\eta = q_{av}/q_{in}$$

Actually, the collector's efficiency is an index of the device's capacity to exploit the available solar source to meet the users' requirements. The higher the collector's efficiency, the larger is the percentage of usable received energy [5].

Efficiency depends on:

- temperature and radiation outer conditions: when room temperature decreases, the collector holds back the heat with more difficulty;
- thermal vector fluid temperature: the greater the difference between the temperature of the pipes and that of the fluid, the quicker and the more efficient is the heat exchange;
- the structural characteristics of the collector: the materials chosen, the optical characteristics of the covering, the absorber plate, the kind of connection between the plate and the pipes, all of which are indicate that the collector's efficiency depends on the its ability to restrict the different outward losses that are always present.

Since the analytical formula for the collector's efficiency is very complex, constructors, installers and design engineers tend to use the practical formula:

$$\eta = A - B \cdot \Delta T^* \quad \Delta T^* = (T_{fm} - T_a)/G$$

where A is the factor which, given that it is constant, sums up the material's optical characteristics and represents the maximum radiation power which the fluid can actually reach; B represents the collector's ability to hold back the acquired/received heat; as for A it is given that B has a constant value; G is the global radiation received on 1 m^2 of intercepting surface in a time unit (W/m^2); T_{fm} is the average temperature of the thermal fluid which flows inside the collector; T_a is the ambient temperature.

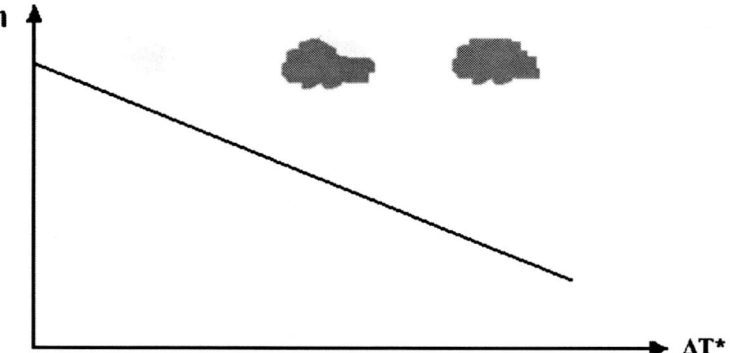

Figure 12: The efficiency curve for a flat plate solar collector.

To guarantee a rigorous comparison between two devices, the collectors' efficiency curve is one of the documents that should be certified by the application of ISO 9806-1 standards (glazed collectors) and those of ISO 9806-3 (uncovered collectors, which will be analysed later). The simplified formula given above allows us to easily represent the efficiency curve, which then becomes a simple straight line (Fig. 12) [1, 5, 9].

Once we have described the different kinds of collectors available on the market, we will be able both to compare the different efficiency curves and to understand their meanings. However, by now it is only possible to compare the efficiency of a flat panel with a copper absorber lined by TiNOX selective materials with the one of a flat panel with a black painted copper absorber (Fig. 13).

The graph in Fig. 13 shows the efficiencies of the two collectors (blue line for the selective collector and black line for the non-selective collector) while the outside temperature decreases. If during the warmer periods of the year the two efficiencies almost coincide, during the colder period the efficiency of the selective collector's is nearly three times higher than that of the black painted collector [18].

Hydraulic connection plans for solar panels The most used hydraulic connection plans for solar panels are shown in Fig. 14.

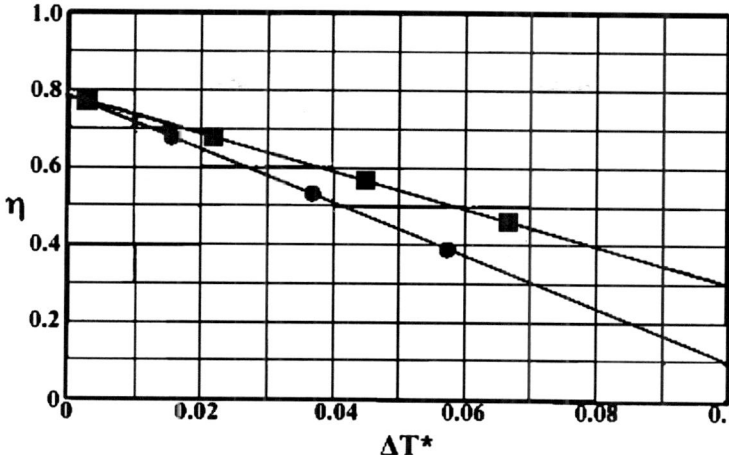

Figure 13: Comparison between the efficiency of a flat panel with a copper absorber lined by TiNOX selective materials (line with squares) and that of a flat panel with a black painted copper absorber (line with circles).

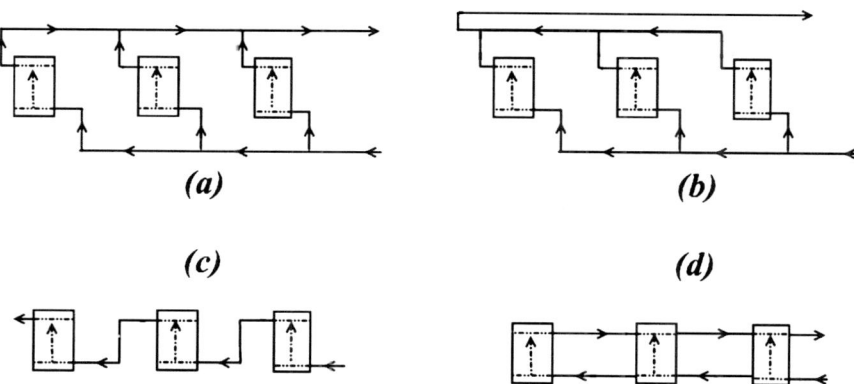

Figure 14: Hydraulic connections for solar panels: (a) parallel; (b) parallel with inverse return; (c) series, to a maximum of five panels; (d) series/parallel, to a maximum of five panels.

These connection systems can be described as follows [2]:

- in the parallel connection, panels work with equal sending and return temperatures;
- in the parallel with inverse return connection, longer pipes are required;
- in the series connection, temperature gradually increases and load losses are bigger, which results in a higher final temperature;
- the series/parallel connection is the cheapest as regards its realization, so it is generally used in small solar systems that have only two panels.

Example of solar panel which is currently commercially available We will now analyse some of its features to show the best solutions [18]:

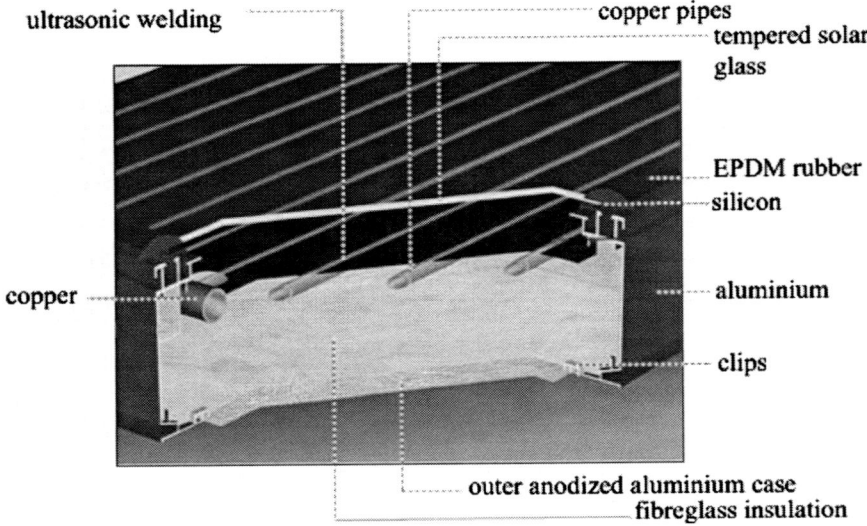

Figure 15: Section of a flat solar panel which is currently on the market.

- *Outer anodized aluminium case*, possibly black coloured, it is aesthetically very attractive and it has great corrosion strength that lasts a few years.
- *Back and side fibreglass insulation* of no less than 4 cm thickness, at least as regards the back part, while the upper part is lined by an aluminium sheet; an excellent absorber insulation to the ambient.
- *EPDM rubber* (thermal polymer ethylene/propylene/diene) *and silicon contour* to guarantee the waterproofing of the upper part which is more exposed to the rain.
- *Low iron tempered solar glass* to avoid danger if it accidentally breaks and to guarantee the best transparency to solar rays.
- *Black painted copper pipes* to get good solar energy absorption even from the elements which connect the selective absorber to the rest of the solar system.
- *Ultrasonic welding* to guarantee good thermal energy transfer from the absorber to the pipes through which the water that has to be heated flows.
- *A TiNOX coated copper thin absorber* to obtain the best conversion of almost all the solar ray frequencies into heat and to decrease the reflected light to its maximum as otherwise it would not be exploited.
- At least a 5-year warranty on manufacturing defects.
- *A European agency certification* which certifies its production, the quality of its materials and the producer's earnestness.

The glazed flat collector is the most common and well known on the market because of its versatility. It can be used in different ways and in different working conditions. As regards sanitary hot water production, this collector is the most used. This collector is know for its excellent cost/performance ratio; it lasts at least 20 years and is capable of supplying hot water from 30°C to 90°C [1, 2, 5, 17].

Working conditions:

- Every kind of place (latitudes and altitudes)
- Any time of the year

Possible uses:

- Small solar systems installed to produce sanitary hot water for domestic uses
- Medium–large systems installed for domestic heating or to produce sanitary hot water for several users
- Solar systems installed to produce low temperature heat for industrial use

To meet the requirements of some specific users or to manage in the climatic conditions of certain places, specific kinds of collectors have been put on the market. We will now describe these in the next few paragraphs.

2.2.1.2 Unglazed collectors Unglazed collectors are widely diffuse and are known for their hot season uses. These collectors do not have a transparent covering, an outside insulation and a container case. In these collectors, the absorber body is made up of a sum of tubes which could be obtained by the extrusion of some plastic materials (polypropylene, neoprene or PVC). The thermal vector fluid flows in these tubes. These tubes are produced in a modular sheaf, which is one metre wide and has a variable length, and are connected at their ends with pipes of the same materials as the collector tubes. The collector is only composed of the absorber plate [4, 5].

Figure 16: Unglazed collector.

They have a much lower cost than the glazed panels and their installation is so simple that it can be done without the help of skilled workers.

To work well these collectors need high outside temperatures and can warm water from 10°C to 40°C, depending on their model. For this reason, these collectors are cut out for hot season uses (bathing establishments, seasonal hotels, camping, holiday houses, etc.).

Figure 17: Installed unglazed collectors.

The use of unglazed collectors is restricted to applications which do not require high temperatures. Since there is neither a glazed covering nor a thermal insulation, in case of high temperature applications the heat losses would be too big, while the efficiency would be too small. These collectors can be installed on flat roofs or on pitches. A wind protection would increase their efficiency. They last nearly 30 years [5, 9, 17].

Working conditions:

- Temperate climates
- Only hot season

Possible applications:

- To warm up water in open air swimming pools;
- In systems installed for hot season use (we talk about bathing establishments, camping, seasonal residences).

Figure 18: Unglazed panels used to warm up a swimming pool.

2.2.1.3 Vacuum tube collectors Although vacuum tube collectors (or evacuated tube collectors) are the most sophisticated and expensive technology, they allow the use of solar systems utilization during all 12 months of the year, even in a harsh climate.

Figure 19: Vacuum tube collectors.

To restrict heat dispersions which are typical in a collector and to improve the efficiency in the vacuum tube collectors, a vacuum is created between the glazed covering and the absorber plate. To completely eliminate the thermal dispersions by convection, a vacuum is created inside the tubes until a pressure less than 10^{-2} bar is obtained. A stronger evacuation allows avoiding the losses caused by thermal conduction. The losses due to radiation, as in the case of flat plate collectors, may be reduced by treating the absorber plate with selective materials. So in these ways, heat losses are considerably reduced, and even when the temperature of the absorber plate is more than 120°C, the outer surface of the tube is cold to the touch. The most important feature of the vacuum tubes which are on the Italian market is that they reach a pressure of 10^{-3} bar (in a few cases, air could be sucked up until a pressure of 10^{-5} bar is achieved) [9].

There are different constructive typologies for evacuated tube collectors. In particular, evacuated tube collectors can have a flat or curved metallic sheet which crosses the glass tube horizontally and works as an absorber plate or a selective coating left on a glass bulb which is inserted inside the glass tube where the vacuum is created. Moreover, a constructive typology does exist, characterized by an absorber plate which is a metallic cylinder put inside the two glass tubes between which the vacuum is created. Its tubular structure is particularly good to balance the stress caused by the atmospheric pressure on the outer surface of the tube [9].

An evacuated tube collector consists of a row of parallel tubes which are joined by connecting their upper ends to a gathering pipe through which the thermal vector fluid flows. The tubes are then fixed by connecting their lower ends to a fitting support.

The evacuated tube collectors currently on the market are [9]:

- *Evacuated tube collectors with a direct circulation system:* The thermal vector fluid absorbs the heat directly circulating inside the vacuum tubes.
- *Heat pipe evacuated tube collectors:* The thermal vector fluid only circulates inside the gathering pipe, which connects the vacuum tubes, without entering the tubes. Each vacuum tube has a second fluid which, during its passage inside the tube, evaporates and then transfers its heat to the thermal vector fluid by condensation.

Evacuated tube collectors with a direct circulation system This constructive typology includes two different solutions. As regards the first solution (a), for which two different current applications are shown in Figs 20 and 21, the solar system consists of two coaxial tubes through which the thermal vector fluid flows. This fluid initially flows inside the inner cylinder, then when it arrives at the base of the glass bulb, it reverses its course and circulates inside the air space between the two coaxial tubes. The second solution (b) (Fig. 22) has a little pipe which longitudinally crosses the metallic cylinder, which works as an absorber plate, following a U-shaped route.

Evacuated tube collectors with a direct circulation system can be oriented southward at their optimum inclination with respect to the latitude of a place. Moreover, by virtue of the absorber plate's curvature, they can be exposed horizontally too.

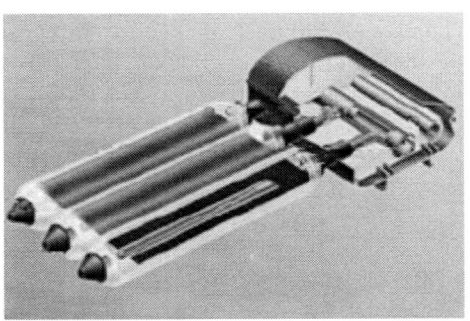

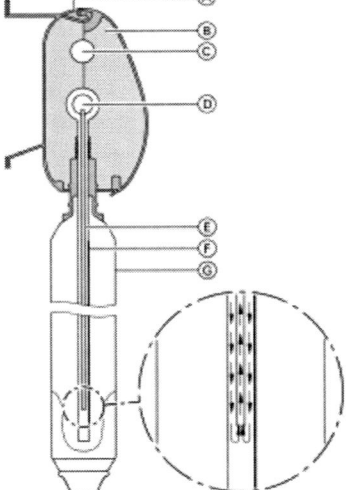

(A) box connection
(B) thermal insulation
(C) pipe flow
(D) coaxial tube collector
(E) coaxial tube heat exchanger
(F) absorber
(G) glass vacuum tubes

Figure 20: Example of an evacuated tube collector with a direct circulation system (solution (a)).

Figure 21: Another example of an evacuated tube collector with a direct circulation system (solution (a))

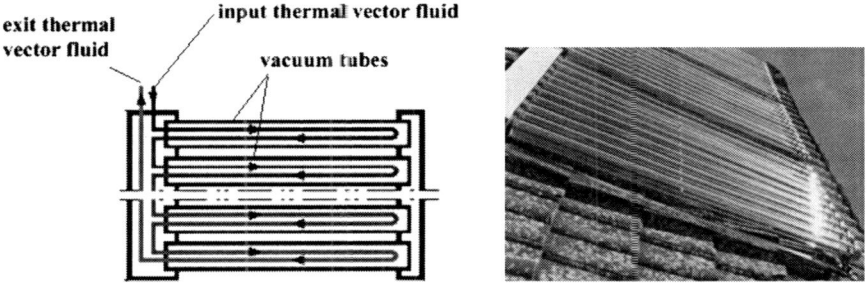

Figure 22: An evacuated tube collector with a direct circulation system (solution (b)).

The Sydney type collector (Fig. 23) is a kind of evacuated tube collector with a direct circulation system. This collector consists of two coaxial glass tubes between which a vacuum is created. The inner tube is covered by a copper sheet which is treated with carbon selective material. Inside that tube, there is an opportunely shaped thin sheet which favours thermal conduction between the absorber plate and the U-shaped pipe through which the thermal vector fluid flows.

This kind of collector generally has a variable number of tubes, which depends on the supplier (from a minimum of 6 to a maximum of 21 evacuated tubes). To increase its capacity to intercept solar radiation, the collector has a few reflectors which are suitable for installation on steep roofs. The flat roof mode does not have these reflectors and so it is better to install this system on a reflecting surface (e.g. gravel).

However, one of Schott's collector models (Figs. 21 and 24) consists of three different glass tubes fitted one in another: the outer tube works as the absorber

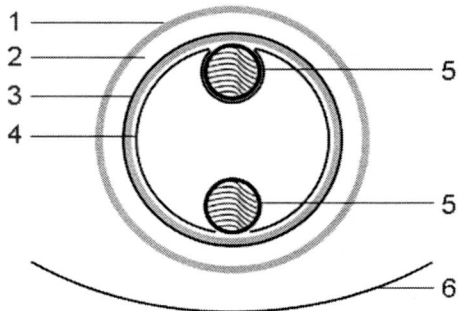

1) Outer Tube
2) Vacuum
3) Selective copper thin sheet
4) Heat conduction thin sheet
5) U-shaped tube
6) Reflector

Figure 23: Sydney's tube scheme.

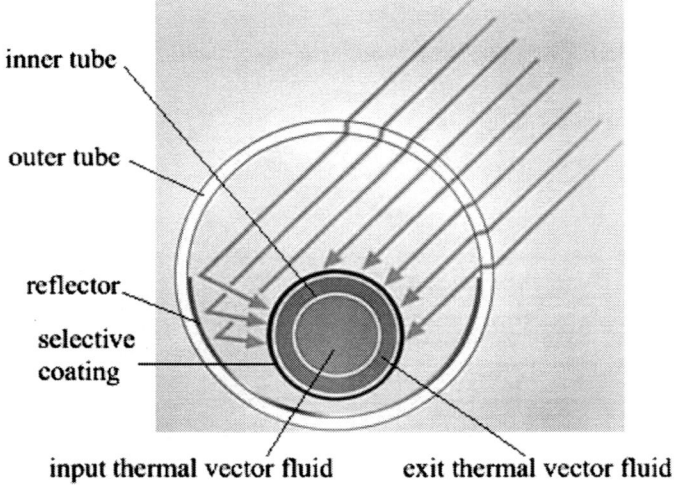

Figure 24: Schott's evacuated tube collector.

plate and it is covered by a selective treatment and the inner tube. The thermal vector fluid flows inside the inner tube; once it reaches the base of the glass bulb, it reverses its course and warms up, circulating inside the air space between the two coaxial tubes.

The vacuum zone pressure, where air has been sucked up and an inert gas has been put in, reaches 30 mbar. A silver coloured reflecting surface is longitudinally put on the lower half of the outer tube, putting the reflector inside the outer tube, so the protection against outside agents increases. [1, 5, 9, 17].

Heat pipe evacuated tube collectors In this kind of device, the heat is exchanged by the passage of phase: the fluid, which is warmed by radiation, first evaporates and then condenses when it comes into contact with a condenser and gives back the thermal energy which it previously absorbed.

The evacuated tube has a thin flat sheet inside which is placed longitudinally and treated in a selective way. This thin flat sheet works as an absorber plate and on it there is a vacuum pipe (heat pipe) which receives the heat from the absorber plate by conduction. The heat pipe is closed and generally contains water or alcohol which evaporates at about 25°C in vacuum conditions. The vapour goes up to the collector's head and there it transfers its thermal energy to the thermal vector fluid which flows inside the gathering pipe. At this point, a new thin liquid film is created which comes back into the evaporation zone by gravity (see Fig. 25). To work properly, tubes have to be installed at an inclination of more than 25°.

This kind of collector has two models currently on the market: a model with a dry heat exchange and another with an immersion heat exchange.

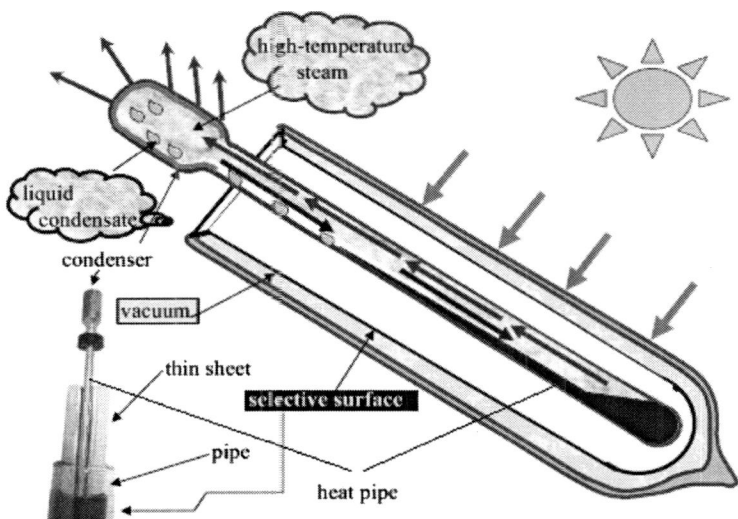

Figure 25: Working scheme of a heat pipe evacuated tube collector.

In the first case (Fig. 26), the thermal vector fluid, which receives the thermal energy produced by condensation from the vapour, flows in a separate pipe. This pipe surrounds the condensers which are placed in series, so the thermal flux flows along the metal walls of this exchanger-pipe (Fig. 27). This constructive solution allows us to easily substitute one of the collector's tubes when it breaks without having to empty out the solar circuit (the connection between the collectors and the energy tank).

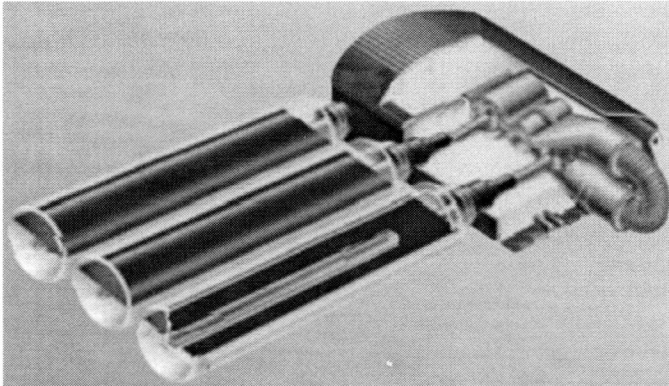

Figure 26: Transverse section of a heat pipe evacuated tube collector with dry heat.

Figure 27: Heat exchanger (dry exchange).

In the second collector model (Figs 28–30), i.e. with an immersion heat exchange, the condensers are directly dipped in the thermal vector fluid. If a tube breaks, we need to empty at least the thermal vector fluid gathering pipe, which is in the head with the vacuum tubes [1, 5, 9, 17].

The efficiency of an evacuated tube collector is on average higher than that of a flat plate collector by virtue of the reduction of thermal dispersions which could be obtained by this system.

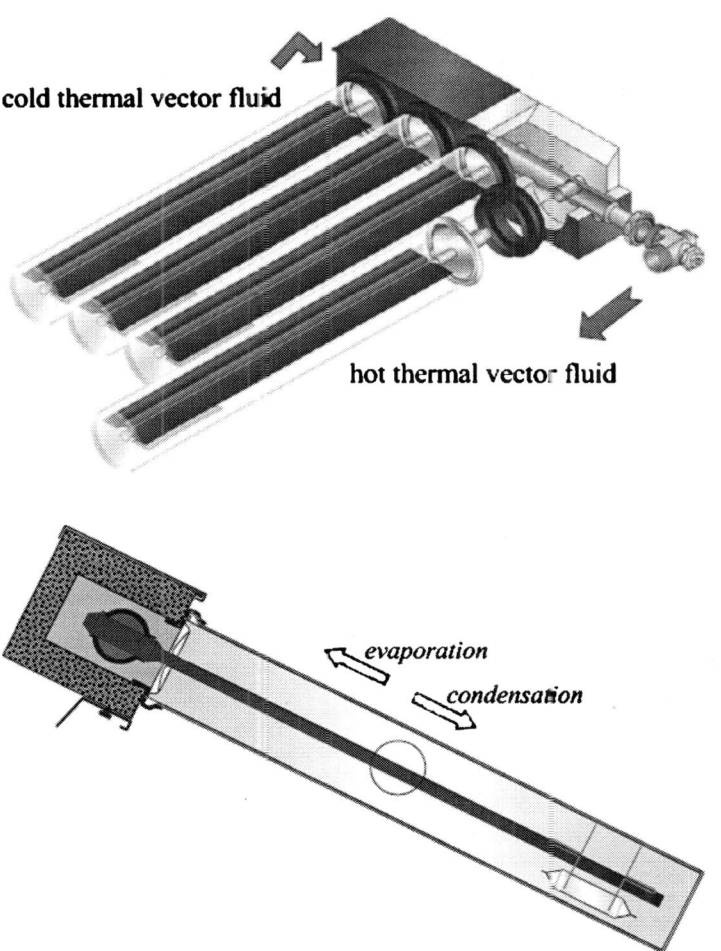

Figure 28: Working scheme of a heat pipe evacuated tube collector with immersion heat exchange.

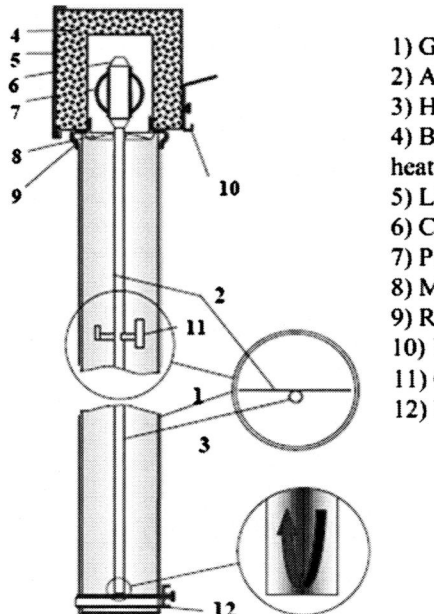

1) Glass tube
2) Absorber
3) Heat pipe
4) Box collector with thermal insulation heat-resistant
5) Lid of the box collector
6) Condenser
7) Pipe with bulb immersion
8) Metal cap
9) Ring seal of the box
10) Upper guide
11) Getter to Bario
12) Lower guide

Figure 29: Components of a heat pipe evacuated tube collector with immersion heat exchange.

Figure 30: Transverse section of a heat pipe evacuated tube collector with immersion heat exchange.

The advantages of an evacuated tube collector are [9]:

- It has excellent efficiency even when the temperature differences between the absorber plate and the outside are very high.
- It has a high efficiency even in case of reduced radiation conditions (e.g. in winter).
- It allows heating of the thermal vector fluid to high temperatures and so it can be used in heating systems, in room conditioning systems and also in vapour/steam production.
- It can be easily transported to any place where it has to be installed.
- It can be oriented southward easily, even during the assemblage phase (this only refers to some products). In fact, in this phase, the tubes can be turned round to place the thin sheets, which work as the absorber plate, perpendicular to the direction of the solar radiation.
- As regards the collector with a direct circulation system, it can be directly installed on a flat roof, reducing anchorage problems and, of course, installation costs.

The limits are:

- It is more expensive than a flat plane collector.
- The heat pipes have to be installed at an inclination of more than 25°.

Working conditions

- Places with low outside temperature or short solar radiation
- All the times of the year

Possible applications:

- Heat production at a higher temperature (e.g. industrial process heat)
- To heat sanitary water or rooms

2.2.1.4 Integrated storage collectors Integrated storage collectors are not very common, although in suitable working conditions they could offer more advantages than flat plate collectors. They could be an easy and interesting solution especially in places with a mild climate. These devices consist of only a single element which substitutes the absorber plate, the serpentine and the hot water storage tank. They are, for example, a series of flanked pipes which have a diameter of 10 cm, a group of pipes of similar dimensions which are created by placing two sheets face to face, a unique big pipe or a variable form tank; the most important thing is that the element in question is able to substitute the storage tank which is generally placed outside the collector [5, 9].

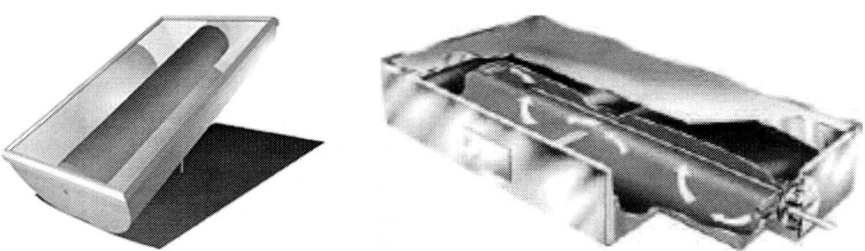

Figure 31: Integrated storage collectors (Figure at left reproduced with permission from Schweyher Engineering, www.schweyehr.com).

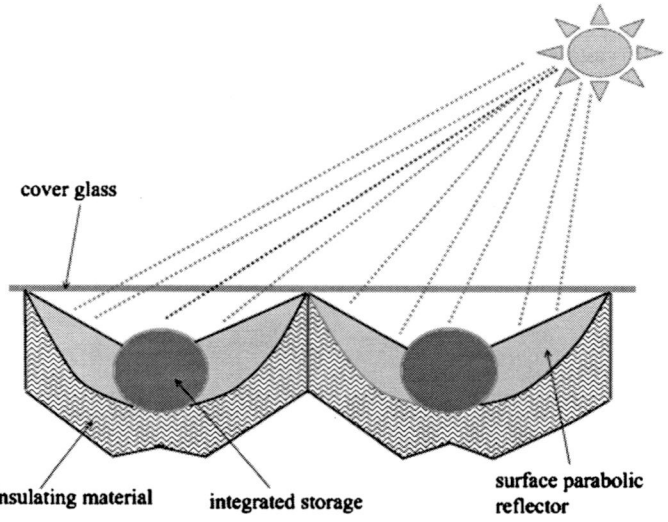

Figure 32: Scheme of an integrated storage collector.

The water, which stands inside the collector and will be used later, absorbs the heat and uniformly diffuses it inside the collector, thanks to spontaneous convective motions. To get an idea about the differences between this collector model and the traditional models, it is sufficient to underline that the water storage per square metre of an integrated storage collector's intercepting surface reaches 80–100 l/m^2 while the storage for devices with an outer storage tank reaches 0.6–2 l/m^2.

However, this kind of collector has the problem of not being able to restrict the outward heat dispersions; actually, only five of the six storage tank's surfaces which are exposed to weather conditions are insulated. Therefore, it is clear that when weather conditions get worse, the integrated storage collector's efficiency quickly decreases, indicating that these devices are unsuitable for sufficiently mild climates or periods of the year [5, 9].

Figure 33: Integrated storage collector.

In comparison with other collector typologies, the integrated storage collector is cheaper, compact, handy, occupies lesser room and can be installed without the help of skilled workers.

Working conditions:

- Temperate climates
- Mild periods of the year

Possible uses:

- Sanitary water heating
- Water heating for low temperature industrial processes

2.2.1.5 Spherical collectors Some thermal solar solutions could regard the system's shape or aesthetic impact. In this case, a spherical system could be less invasive and cheaper. It is a simple integrated storage collector [17, 25].

2.2.1.6 Air collectors This is another kind of collector suitable for low temperature applications. They are very similar to normal glazed panels, but in this case air is used as the vector fluid rather than water. Air can circulate between the glass and the absorber or between the absorber and the bottom of the panel. Usually, the absorber is finned to make the passage of air slower and tortuous, because air exchanges heat with more difficulty compared with water. So, we require that the air remains inside the panels longer to make it absorb the greatest quantity of heat. Since air never freezes, there is no need to use antifreeze techniques. Currently, these collectors have reached 60–70% efficiency and they have a long life-time (even over 30 years).

Figure 34: Spherical collector.

Figure 35: Air collector.

Air collectors may have different applications such as heating water or producing compost using toilet emissions. However, their principal application is in solar heating for buildings. In fact, one particular example of air solar panels is the lining panels which are used as a coating for normal plugging walls in industrial, commercial and residential buildings. They are not glazed but have an outer metal surface which works as an absorber and heats the air which enters the collector through micro-perforations. The air which circulates inside the air space between the panel and the wall can then be circulated inside the rooms using a proper aspiration system, thus contributing to heating and changing of air in the same rooms. During the summer, they help to bring down the temperature by not allowing the solar radiation to fall directly on the building's external walls. When the fan is switched off, fresh air enters from the lower perforations and by natural convective motion it goes out from the higher perforations, creating a continuous flow/flux which helps to maintain the wall's temperature.

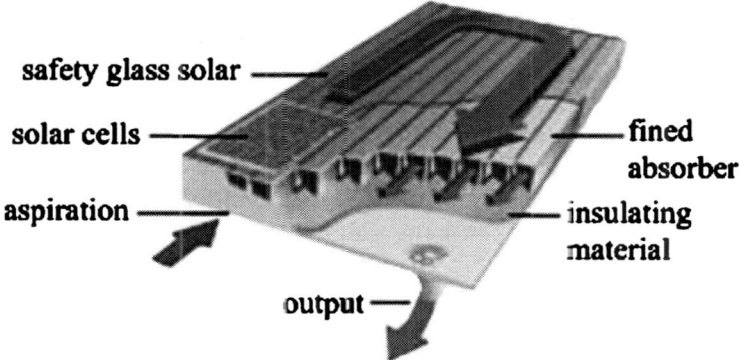

Figure 36: Working scheme of an air collector.

These systems are able to supply between 25% and 50% of the energy needed to heat a room [16, 17, 23, 25].

2.2.1.7 Efficiency curves in comparison Figure 37 shows the efficiency curves of the most diffuse collector typologies (see section 'Collector's efficiency' under par. 2.2.1.1). These collectors are the flat plate collector (with and without a selective treatment), the unglazed collector and the evacuated tube collector. Taking into consideration the fact that they have the same thermal vector fluid temperature, the increasing in the ΔT^* value concerns the unfavourable weather conditions (radiation and/or decreasing ambient temperature). Therefore, moving rightward along the abscissa we can see how the collector's efficiency changes, if working conditions get worse [5].

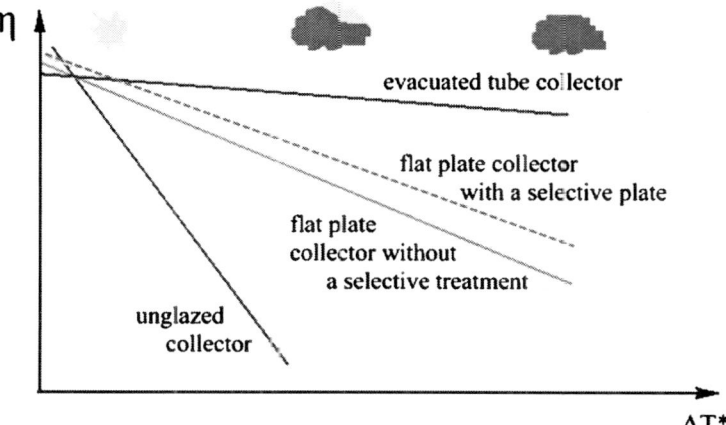

Figure 37: Comparison of the efficiency curves of different kinds of collectors.

From the efficiency curves, we can immediately observe the differences between these kinds of collectors [5, 9]:

- Since the unglazed collectors does not have a transparent covering, it has the best chance to absorb the incident radiation (the straight line meets the ordinate axis at the highest point); however, its efficiency decreases rapidly until it reaches zero in places where the other kinds of collectors have a decent efficiency.
- The flat plate collector with a selective plate has a better performance in every working condition compared with the simple flat plate collector.
- The evacuated tube collector has the most stable efficiency curve and it guarantees good performances even during unfavourable working conditions.

As an example, Table 2 lists some values of the A and B coefficients for some of the commercial devices available in the Italian market.

Table 2: A and B coefficient values.

	A	B
Flat plate collector without a selective treatment	0.70–0.85	5.5–7.6
Flat plate collector with a selective plate	0.75–0.85	3.5–5.8
Unglazed collector	0.80–0.86	22.0–28.0
Evacuated tube collector	0.80–0.85	2.0–3.0

2.2.2 Typologies of solar systems

The passage from the collector to the solar system requires few elements which can make the service enjoyable to users and stabilize the collector's performance [5].

- The role of the *storage tank* is to supply hot water to users at any time and in any weather condition, whenever it is required; it stores the water heated by collectors in small amounts and maintains the hot water at a constant temperature until it is demanded by the users.
- The *auxiliary system* (specifically methane central-heating boiler or electric water heater) is needed to make up for any contingency in the solar source and for the lower solar energy availability during winter. In this way, we avoid oversizing the solar system until it becomes too expensive.
- The *expansion tube* is the part which is able to receive any excessive thermal expansion of the thermal vector fluid, thus avoiding the creation of dangerous overpressures.
- *Safety-bolts and system check* ('jolly' valves, intercepting valve, thermostats, etc.)

We can find other circuit elements (such as circulators, control station, etc.) only in a few kinds of system.

2.2.2.1 Nomenclature and principal applications Solar energy systems can be divided into four principal categories: first, according to the relationship between the thermal vector fluid and the service given to users; second, according to the way in which the fluid is circulated.

We talk about *open systems* when the fluid inside the collector is water itself, which is provided to the users once it has reached the required temperature. By contrast, we talk about *closed systems* if the thermal vector fluid flowing inside the collectors transfers its heat to the usable fluid (water) through an exchanger. In the latter case, we have two distinct circuits, one for the thermal vector fluid and another for the water that has to be heated up [5, 6, 17].

Observing the circulation of the thermal vector fluid inside a solar system, we must distinguish [5, 6]:

- *Forced circulation system:* In this case, to regulate the flux, we need to insert an automatic system which consists of a circulator with thermostats and a control station.
- *Natural circulation system:* In this case, the fluid's flow inside the collector is automatically stabilized by spontaneous convective motions.

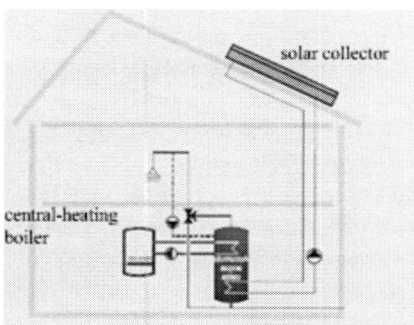

Figure 38: Forced circulation system and an example of the installation of a forced circulation system.

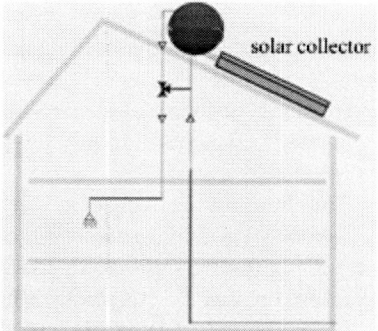

Figure 39: Natural circulation system and an example of the installation of a natural circulation system.

In theory, the two variables are completely independent and so it is possible to install systems having all the four possible combinations. However, the experience acquired through installations has shown that only a few possible solutions can be put into practice. The advantages of an open circuit system are the simplicity of its hydraulic circuit realization and the absence of thermal dispersions, which occur every time heat moves from one circuit to another [5].

Nevertheless, an open circuit system is not usually adopted for two reasons: (1) water can easily freeze when the temperature is below zero and (2) there could be lime scale accumulation inside the collector's pipes. In both cases, the collector can be damaged till it becomes out of service. Because of these problems, which are difficult to control, the open circuit system has been substituted with more complex installations; however, the simplicity of the open circuit system is still exploited in these cases [5]:

- Systems whose collectors are unglazed and are used only during the hot season: These systems avoid the freezing problem and the lime scale accumulation is limited to the working temperature (not above 40–45°C).
- Integrated storage systems installed in places with a mild climate: These devices use such large quantities of water that would hardly freeze completely and, at the same time, the absence of small sized pipes does not make probable scales very dangerous (but they have to be checked in any case using softener filters).

Except for the above-mentioned applications, the closed circuit system represents the widest and the most reliable solution. In this case, two different hydraulic circuits are involved (actually, closed systems are also called *double circuit systems*): the *primary circuit*, where only the thermal vector fluid flows, and the *secondary circuit*, where water, coming from the water network and assigned to users, flows. The thermal vector fluid absorbs the energy from the intercepting plate and then transfers the greatest part of that heat to the water that has to be warmed. The place where the heat transfer takes place is a very important element and it is called *the exchanger*. Since walls with high thermal conductivity separate the thermal vector fluid from the water, the fluid transfers heat to the cold water in proportion to the temperature difference between the two liquids. If the interface surface is large, the thermal energy exchange will also be high, especially when temperature differences may not be relevant. To satisfy the need to have a large exchanging surface and at the same time a compact device to rely on, the most common exchangers currently in use are in the form of a dipped worm-pie, a sheaf of tubes or plates.

The choice of the exchanger is really important since a good performance of this device, besides making the service quicker and more efficacious, allows the thermal vector fluid to return to the collectors at a highly decreased temperature, which in turn increases the collectors' efficiency [5].

As we have just examined the differences between open and closed systems, we go on to analyse the circulation of the thermal vector fluid (either water or anti-freeze solution) inside the system.

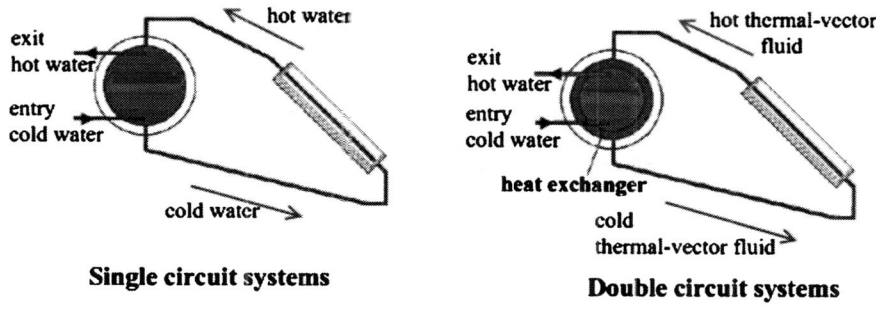

Single circuit systems   **Double circuit systems**

Figure 40: Working scheme of a natural circulation system.

Natural circulation [5, 6, 9] (Fig. 40) exploits the spontaneous behaviour of fluids to create convective motions when there is a localized increase in the temperature. The systems which exploit this phenomenon can be realized using any kind of solar panel and are characterized by a storage tank which is elevated compared to the collector (Fig. 41); thanks to this property, the fluid in the collector, once it has heated up on coming into contact with the exposed plate, becomes less and less thick and spontaneously tends to move up towards the storage tank. In this way, it leaves enough room for the fresher fluid inside the collector.

Figure 41: Natural circulation.

Therefore, to obtain excellent performance, this kind of device regulates itself by optimizing the fluid circulation spontaneously. However, the system's structure has some characteristics that restrain its utilization; for example, the storage tank is completely exposed to all weathers and seasonal variations. So, even if the storage tank is correctly and efficaciously insulated, it cannot avoid the host of energy losses when exposed to very low temperatures.

Second, since the storage tank is located above the collector, the system can have remarkably high weights especially when the intercepting surfaces are very big. In this case, the system's weight can become a problem for the structural resistance of the roof and garret, as these are the usual places of installation.

Eventually, the aesthetic impact of the most common natural circulation collectors is not one of the best: their structures make them particularly showy (Fig. 42). To overcome this problem, scientists have been realized a few less showy devices which have the storage tank behind the collector, for example (Fig. 42).

Figure 42: Natural circulation collectors with storage tanks above and behind the panel.

If the roof is steep, the storage tank can be installed inside the building, although it stays above the collector. In this case, we will not have any energy losses or aesthetic impact problems.

However, as regards a small dimension system which is installed in a place with a mild climate, the natural circulation remains the best solution as it is simple, cheap and compact.

Natural circulation may be used in the following cases [5]:

- integrated storage solar collectors which are placed in any manner;
- monobloc solar collectors (i.e. with a storage tank fastened to the upper part of the collector);
- solar collectors installed on the ground and the storage tank (separated from the collectors) located on an elevated structure which is inside the building;
- solar collectors installed on the roof's slope and the storage tank (separated from the collectors) placed inside the garret and located in a more elevated position than the collectors.

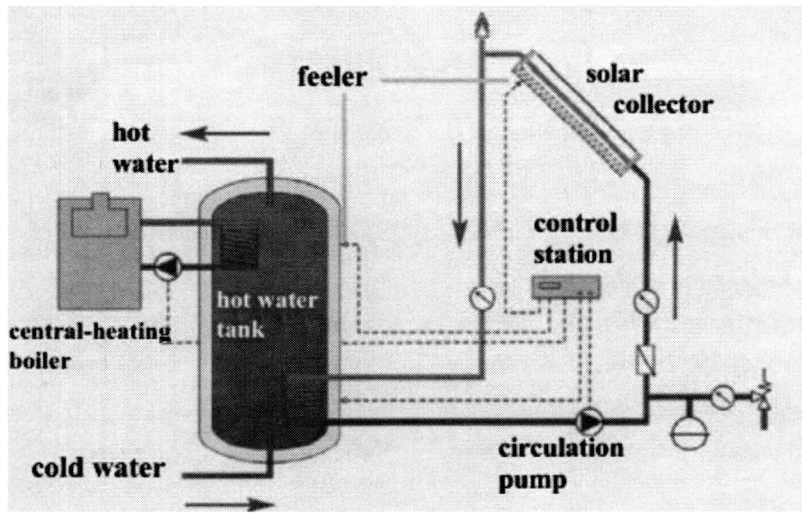

Figure 43: Scheme of a forced circulation system.

Forced circulation [5, 6, 9] is always necessary when it is not possible to place the storage tank in a higher position than the collector. In this case, the best circulation for the thermal vector fluid would be one which is completely opposite to the circulation which is considered natural. Because of this, the circuit needs a few additional devices such as a *circulation pump*, which moves the fluid in the right direction, a *non-return valve*, which does not allow the reverse circulation to take place in any situation, and a *control station*, which automatically operates inside the circulator to regulate the fluid circulation and optimize the system's performance.

This system is certainly complex and expensive and also requires that each of its parts is accurately proportioned. However, it gives us freedom in terms of its design and architectural integration (the storage tank is actually completely separated from the collectors) and is also suitable for any weather condition.

Forced circulation systems can be realized using any kind of collector, excepting the integrated storage collector, since this collector also works as a storage tank.

We are almost forced to use a forced circulation system [5]:

- when we want the system to be more precisely checked and regulated;
- when the weight of the storage is more than the roof's resistance;
- when there is no garret where we can install the storage tank, and its aesthetic impact poses a serious disadvantage;
- when, due to logistic reasons, it is not possible to realize a natural circulation.

Figure 44: Images of forced circulation systems.

The advantages of natural circulation systems are:

- the speed of thermal exchange is proportional to the temperature difference between the storage boiler and the panels;
- the circulation is self-regulated;
- there are no circulation pumps, control stations and feelers;
- quick and cheap installation;
- minimum maintenance.

The advantages of forced circulation systems are:

- architectural integration of the collectors;
- maximum flexibility of the system.

Table 3 lists the principal applications of each system in terms of its typology.

Table 3: System typologies used in various applications.

Circulation	System's typology	Principal applications
Natural	Open system	Small systems used to heat sanitary water (no rigorous climate or with integrated storage collectors) Systems used only during hot season (e.g. bathing establishments or camping)
Natural	Closed system	Systems used to heat sanitary water or to heat rooms
Forced	Open system	Systems used only during hot seasons (bathing establishments or camping) Systems used to heat water in swimming pools
Forced	Closed system	Small systems used to heat sanitary water for domestic use when it is not possible to put the storage tank above the collectors Systems installed to heat sanitary water which will be used by collective users Systems used to heat sanitary water and rooms Systems used to heat water in swimming pools

2.2.2.2 System-type description In this paragraph, we will offer an outline of the most common system schemes as regards the applications listed below [5, 9, 17]:

1. domestic system for sanitary hot water production;
2. big collective use system for sanitary hot water production;
3. small combined systems for sanitary hot water production and room heating;
4. system for heating swimming pools.

Another important application which has to be mentioned is room refreshing during the hot season. Some of the newest and most efficient devices which can be used for air conditioning, i.e. absorption heat pumps, require a hot thermal source to supply air for refreshing to the user. The thermal vector fluid, which flows inside the collectors, can perform this function. This application is particularly interesting since the refreshing system's peak load of work corresponds exactly with the maximum availability of the solar source. Actually, if there is a reduced phase displacement with time, we can assume that the thermal load to be taken from the outside is roughly proportional to the incident solar radiation.

Moreover, this solution is very desirable if we think about the consumption of electricity and air-conditioning costs.

As regards the heat pump which uses helium, the consumption can be limited only to the integration of solar feeding and for this reason its consumption can be lower than that of a normal heat pump. However, this solution is not so common due to the high cost of the absorption heat pumps and the need for reaching higher hot source temperatures. In fact, these temperatures have to be higher than the working temperatures of a flat plate solar collector. Recent studies have underlined how the latter problem can be overcome by the installation of an evacuated tube solar collector furnished with mirrors for the concentration of solar radiation. The economic problem has not been overcome yet, but as the dimensions of the system increase, the more its importance decreases [5].

Domestic systems for sanitary hot water production As regards sanitary hot water production, the most common combinations are:

- natural circulation (or through radiators), closed circuit;
- systems with integrated storage collectors (less common);
- forced circulation, closed circuit.

The first application (Fig. 45) is generally the preferred one.

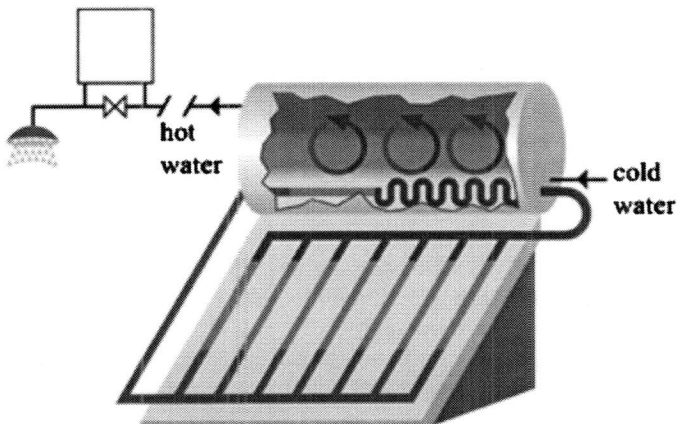

Figure 45: Radiator collector (natural circulation) and closed circuit.

In this case, the primary circuit (where the thermal vector fluid flows) is completely inside the storage–collector system and the exchanger (generally has a dipped worm-pie configuration or an outer cover) is inside the storage tank. As stated before, the regulation of the radiation system is completely spontaneous: the thermal vector fluid, once it has warmed up inside the collector, tends to move up towards the storage where, thanks to the exchanger, it transfers heat to the water which is inside the tank. When the temperature of the fluid which is inside the collector becomes higher than the temperature of the fluid inside the exchanger, because of the density difference, the circulation in the primary circuit begins spontaneously. In this manner, without any outside intervention, the collector is always filled up with the thermal vector fluid which has a lower temperature (which results in the increase of the efficiency of the device) and inside the exchanger there is always the warmer fluid. Since the water which has to receive the heat stratifies upwards when temperatures are high, the exchanger is installed in the lower part of the storage system; in fact, even in this case, it is important that the exchange happens using the colder water which is inside the tank. Some closed circuit radiator collectors have (or are equipped for the insertion of) an electric resistance inside the storage tank, which is switched on by a thermostat when it is not possible to reach the required temperature. The advantage of this solution is to have also, in a single device, the traditional integrative system to supply the required service directly to the users.

The disadvantage instead is the use of electricity to produce heat at a low temperature, although in an integrative and limited way. As we actually know, electricity is a very noble and versatile kind of energy and for its production (in Italy it happens in 80% of the cases) fossil fuels are burnt at very high temperatures, losing almost 67% of the energy developed in the burning phase, i.e. during the thermo-electrical conversions, and during the transport to where the service is supplied.

From an economic, energetic and environmental point of view, the use of electricity for the sanitary hot water production is extremely irrational. So it is preferred, where possible, to use an outer integration such as the gas boiler.

The scheme of an integrated storage collector's system is very similar to the circuit scheme (as the one we have already seen) where a monobloc radiator collector is inserted. If we imagine the collector as a black box, i.e. if we observe only the inputs and the outputs without considering the inner working of the device, the two cases could be superimposable. As regards the integrated storage system, the maintenance and frequent control of the softener filter are very important since the water which is supplied to the users is the same water which passes through the collector and so it is important to reduce the lime scale accumulation which water can cause to its minimum [5].

For all domestic applications where it is not possible to place the storage tank above the collector (or the choice of an integrated storage collector is not convenient), it is very common to choose a forced circulation system with a closed circuit. In this last case, the tank is located in any part of the building and always in a vertical position to favour the stratification of water when the temperature increases and also to reduce the mixing between the cold water that enters and the water that is ready for the users (a phenomenon which has a negative influence on the overall efficiency of the system).

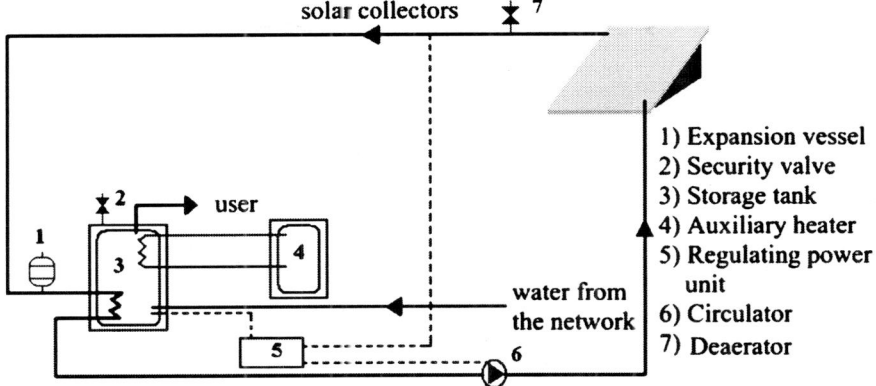

Figure 46: Forced circulation system with closed circuit.

As we can observe from Fig. 46, the primary and secondary circuits as a whole are more complex and articulated compared with the radiator scheme. In this case, it is necessary to use a circulator which is run by a definite regulation power unit.

The circulator is a small centrifugal pump capable of moving the fluid inside the primary circuit. Its running has to allow excellent processing of the collectors (the fluid inside them should not be too hot) and at the same time has to allow efficient heat transmission from the primary and the secondary circuits (the temperature difference between the two circuits should allow the exchange).

To regulate the temperature difference between the two circuits, the circulator has a differential thermostat fitted for measuring the temperature difference between the thermal vector fluid which comes out from the collector and the water which is in the lower part of the storage tank. If the measured difference, for example, is over 5–8°C, the circulator moves the primary fluid, starting the heat exchange between the two circuits; as soon as the measured temperature difference is too small for thermal exchange to take place (going down to <3–4°C), the power unit switches off the pump.

The electrical power unit can also protect the system, having the possibility to control the maximum running temperature of both the collector and the storage tank. The exchanger can be inside or outside the storage tank; this last solution is preferred when the systems are big and normally plate exchangers are installed.

When the hot water storage tank is located in a place which is very near the final users, it results in a small inefficiency due to the presence of cold water inside the tubes fitted for the transfer of hot water from the storage tank. Actually, the users will have to let a few litres of cold water run before getting water at the temperature they require. This problem can be solved by the predisposition of a recycling system inside the general hydraulic circuit; a small quantity of stored hot water periodically passes through the pipes which connect the storage tank to the final user to keep both the tubes and the water at nearly the required temperature. This solution involves a small loss of stored energy but it is the most commonly used technique especially in plants of medium to large size [5].

Big collective use systems for sanitary hot water production In this specific case, we have systems characterized by an absorber plate which exceed an area of 100 m^2; their realization often depends on collective users such as apartment buildings, sport centres, schools, hospitals and hotels.

Figure 47: Pictures of collective use system for sanitary hot water production.

Actually, all the structures which are characterized by a relevant, continuous and concentrated demand for sanitary hot water are interested in this application. The realization of a large-sized system could be favourable in terms of cost: the system's dimensions allow a significant reduction in the total price (because of both the 'stair' effect, which is connected to the components that have to be bought, and the installation costs, which has a lesser influence than the former on the final price).

The most common solution adopted is one characterized by the closed circuit and forced circulation.

In big residential buildings, solar collectors can meet 30–50% of the yearly energy requirement for sanitary hot water. In this case, we talk about 'pre-heating', since a complementary system which brings water to the required temperature is generally needed even during the hot season. The proportioning of the solar system and its combination with the heating system depend on the subject characteristics [5, 17].

Figure 48 shows the scheme of a solar system (with a closed circuit and forced circulation) installed to heat water in an apartment building.

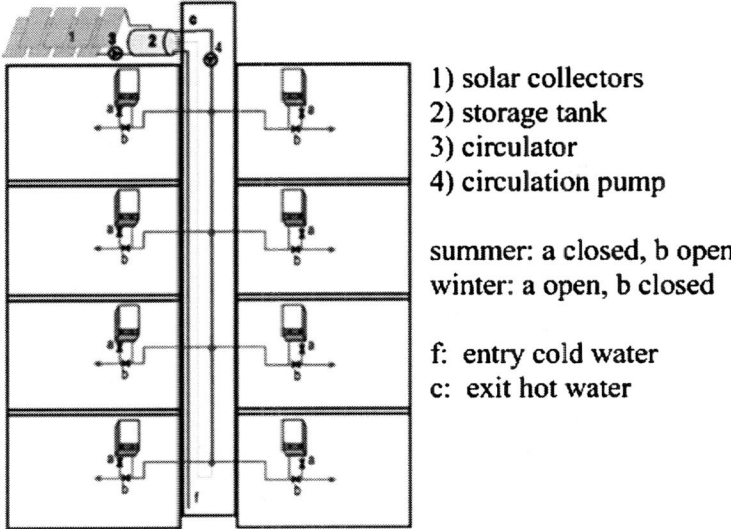

1) solar collectors
2) storage tank
3) circulator
4) circulation pump

summer: a closed, b open
winter: a open, b closed

f: entry cold water
c: exit hot water

Figure 48: Example of a solar system for an apartment building.

The system consists of a boiler (water heater) and a certain number of solar panels which are in proportion to the number of users (the panels must form a receiving surface that has an area of at least 2 m² per standard family of four people). The boiler should be put on the balcony or on the thermal power unit.

Solar radiation warms the fluid which is inside the solar panels. The power unit takes the temperature of the fluid when it goes out from the solar panels and also that of the water inside the boiler. When the thermal fluid temperature exceeds the temperature of the water inside the boiler by a certain ΔT number (which is selected on the electronic power unit), the electronic power unit starts the exchanger and so it gives way to the thermal exchange, warming up the water inside the boiler. This situation lasts until the thermal difference is higher than the ΔT selected on the electrical power unit.

This system self-sufficiently supplies warm water to the users during the spring, summer and autumn months (a closed, b opened), while during the winter it 'preheats' the sanitary water (a opened, b closed), which is heated up later to the required temperature by a gas or electrical central-heating boiler. The advantages of this already tested technique are:

- *Hot water distribution:* The sole ring used to distribute hot water lets all the users use it immediately.
- *Low installation costs for the water-supply:* A single tube is used for cold water feeding and another for warm water feeding.
- *System functionality:* With this system typology, the non-simultaneous utilization of hot water by the users prolongs the self-sufficiency time.

The only cost of this system, which should be shared among the people living in the apartment building, is the electrical energy used by the exchanger, which is equivalent to the energy consumed by a 70 W bulb. The exchanger has to be connected to the electrical system of the stairwell and works for 4 or 5 hours a day. A special subtraction meter, placed in each apartment, allows calculating how much hot water each family consumes [64].

Small combined systems Many studies have indicated that the yearly thermal energy requirement for room heating is two to ten times higher than the yearly sanitary hot water requirement. The idea to move a part of this consumption from the traditional energy sources to the solar source is fascinating: although the requirement from users is more during the part of the year with less solar radiation, the proposed solutions have become interesting even from an economic point of view (which is very important for people living in places with a very rigorous climate).

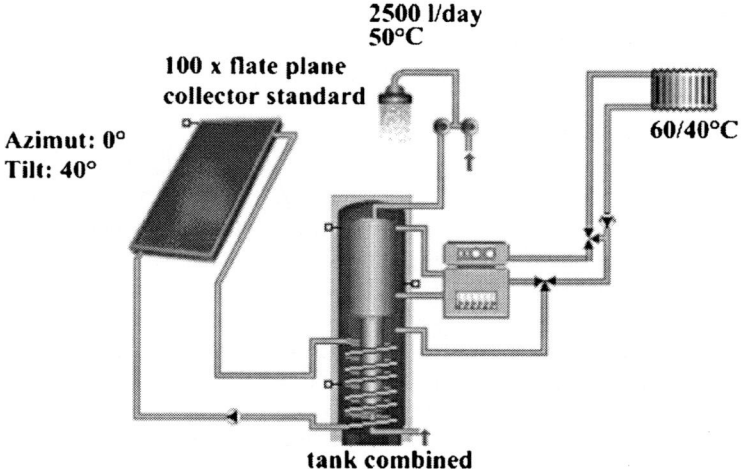

Figure 49: Scheme of a combined solar system (closed cycle and forced circulation).

As regards only room heating, adopting a combined solar system will cover 10–40% of the total yearly thermal requirement. Exceeding these values will be inconvenient from a technical and economic point of view: the big active surfaces which are necessary to supply energy during winter would produce a huge energy surplus in summer, a large part of which would be wasted if there are no particular additional energy requirements.

One solution would be the so-called seasonal energy storage which exploits the heat stored in summer during the cold season. Although this solution sounds very interesting, the technological solutions which rely on this principle are not suitable for small domestic systems from both logistic and economic points of view.

Although they cannot guarantee the self-sufficiency of the system, resorting to daily storage systems shows, on the contrary, a growing improvement in the cost-benefit ratio, even for small systems. Since during winter the best performance of a thermal solar collector can be obtained when the supplying water temperature is about 30–40°C, the room heating system cannot consist of traditional radiators. In fact, traditional radiators do not have a wide exchanging surface, so to heat the rooms it is necessary to feed them water at higher temperature. We may resort to radiant heaters (which are more suitable than radiators for the utilization of solar sources) whose surface extension should be equal, for example, to the entire floor of the house that has to be heated up (see Fig. 50).

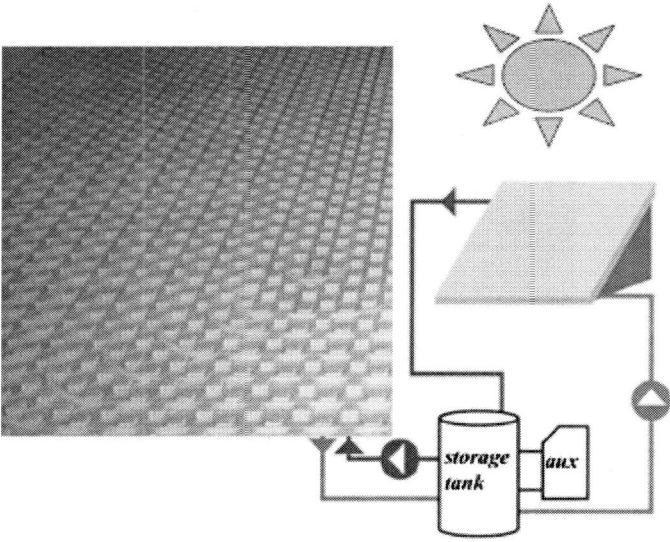

Figure 50: Scheme of a solar source heating system.

As regards radiant devices, it is obviously possible to install different configurations. However, all of them aim to increase the exchanging surface and to guarantee the users' thermal comfort. The radiant heater may coincide with a wall or it can be substituted by particular radiant skirting-boards which run along all the walls of the house.

Based on what we have just mentioned and the difference between the users' requirement and the source's availability, the installation of systems called *combi*, i.e. combined, is interesting if it is associated with specific characteristics of the whole building. First, an efficient thermal insulation is needed to avoid dispersion of the stored heat. Second, it is necessary to have a heating system based on radiant heaters or alternatively to realize a new system together with the restoration of the building to allow the adoption of that system. As stated, the choice of a combined solar system could be particularly favourable if the building has consistent summer users. In this case, the energy surplus produced during the hot months is completely used. This results in advantages in system exploitation and also in terms of economic return of the investment [5, 9].

System for swimming pools The utilization of the solar source for heating water in a swimming pool is fascinating for several reasons: first, the temperature that has to be reached to guarantee the users' well-being is not higher than 25–28°C. This lets the collectors work in favourable conditions as regards the efficiency. Second, it is interesting to observe that the system turns out to be particularly simple, thanks to the possibility of eliminating the storage system, substituted by the water mass contained inside the swimming pool itself (see Figs 51 and 52).

Figure 51: Solar system for heating water in a swimming pool.

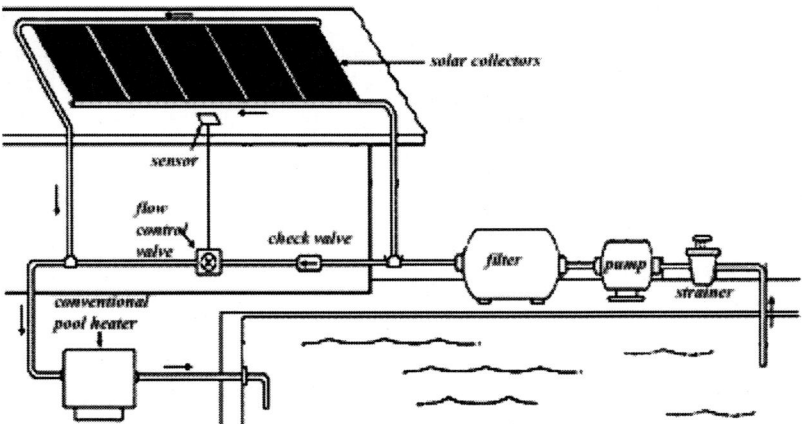

Figure 52: Simplified scheme of a solar system installed to heat water in a swimming pool.

For open air swimming pools, which are mainly used during the summer, it is possible to meet almost 100% of the energy requirement using the solar source. In this case, the high ambient temperatures during the processing of the system, the low thermal vector fluid temperature and the strong radiation allows us to use the unglazed collector, which has been discussed previously (par. 2.2.1.2). The unglazed collectors are characterized by a convenient price (almost 50% less than the price of a glazed panel) and their performance is very appreciable.

More complex systems can provide for, besides heating the water in swimming pools, the requirement of sanitary hot water utilization connected to the showers. As we have already seen in the case of the *combi* systems, in this case also particular care must be taken in the planning and regulation of the thermal priority so as to obtain the highest performances from the system (Fig. 53) [5, 9].

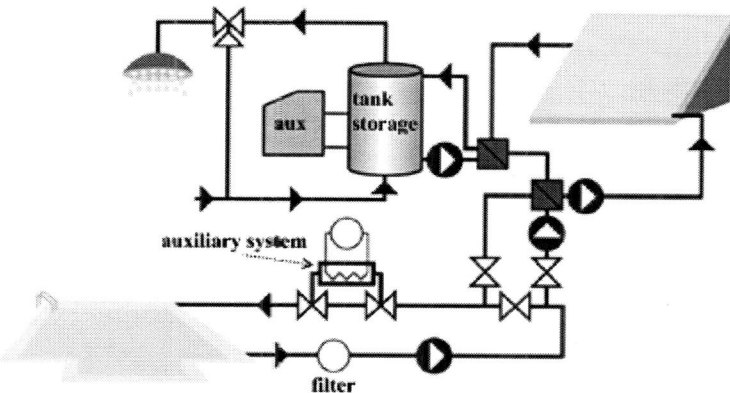

Figure 53: Simplified scheme of a solar system used to heat an open air swimming pool and to produce sanitary hot water.

2.2.3 The solar circuit

The solar circuit is the connection between the solar collectors and the storage tank. A solar circuit can be made of copper or stainless steel pipes. To restrict heat losses, the pipes which connect the collectors to the tank have to be short and insulated to their maximum. Stainless steel pipes have a surface which is less smooth than that of copper pipes and so they cause bigger load loss. In open systems, it is better to use tubes which have a smooth inner surface (e.g. copper) to prevent encrustments. When we choose the circuit components, it is very important to consider the kind of fluid they are going to contain; actually, antifreeze solutions or products used in the swimming pool impose the utilization of components (taps and fittings, sets) which do not corrode when they come into contact with certain chemical substances. The good processing of a solar system strongly depends on how the solar circuit's insulation has been carried out. A sufficient insulation layer and also a good insulation execution without interruptions or escapes are necessary. This also applies to the circuit's elbows. Concerning the choice of the insulating materials, it is important to take into consideration their resistance to high temperatures; for very short periods, the temperature inside the solar circuit's tubes can reach more than 200°C. Moreover, the insulation should be able to resis atmospheric agents and ultraviolet rays. Suitable materials could be insulating mineral fibres or insulating materials such as Aeroflex and Armaflex HT. On the outside, the insulation can be protected by tube-coverings made of a steel or zinc-plated layer.

In the solar circuit, besides the pipes, there are devices which are necessary to guarantee the fluid's motion and the security on the basis of its assemblage and utilization (pumps, buffer vessels, expansion vessels, security valves, air discharge valves, etc.). Generally, all the basic hydraulic components offered are already pre-set by a great majority of the manufacturers. Also, the control instruments (manometers and thermometers) are pre-set [5, 6, 9, 17].

Let us now see the elements which characterize the solar circuit, with reference to systems with a forced circulation, because of its complexity. It has been seen

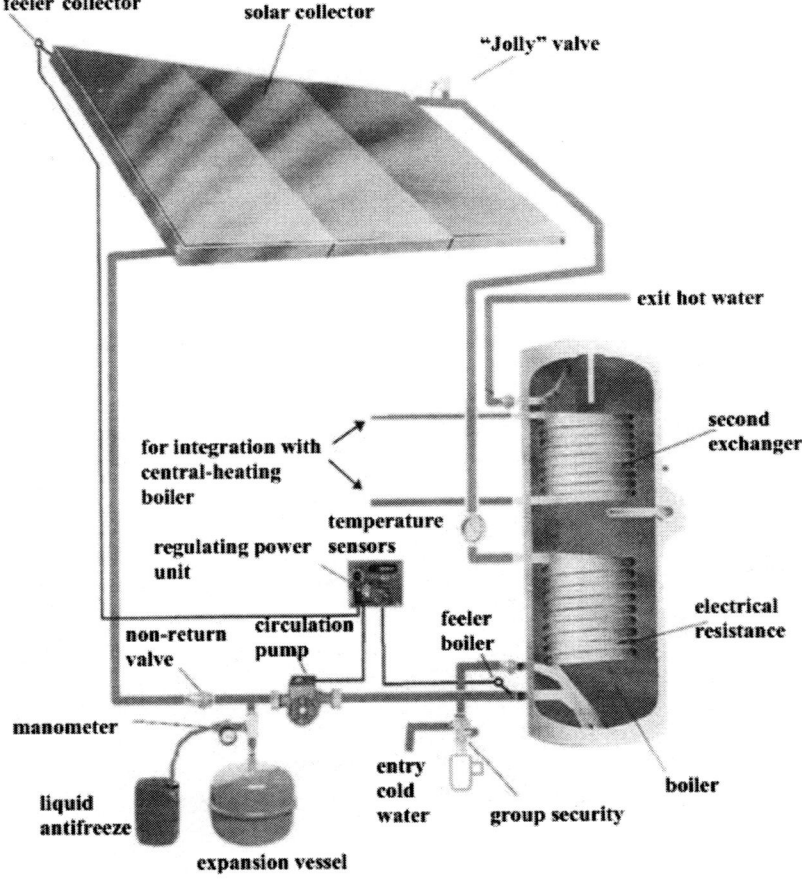

Figure 54: Example of a thermal solar circuit (with a forced circulation).

that in the natural circulation system the storage tank can be directly heated up by the natural circulation or by a heat exchanger. Moreover, there is no device which is able to actively regulate the solar circuit [6].

Pump As in the centralized heating systems, in the solar systems, there are 'lots' and 'returns'. The pipe in which the hot thermal vector fluid flows from the collector to the storage tank is called 'lot', whereas the 'return' is the pipe with the colder fluid which flows from the storage tank to the collector. The pump must be installed on the return line, with the motor's shaft in the horizontal direction. The pump must not be insulated [5, 6, 33].

Figure 55: Circulation pump.

Non-return valve (or restraint valve) When collectors are installed in an elevated position than the storage tank and the fluid inside the tank has a higher temperature than the fluid which flows inside the collectors (especially during the night), the temperature difference between the hot fluid inside the exchanger (in a lower position) and the cold fluid inside the collectors (in a higher position) would start a natural circulation inside the primary circuit causing dispersion of the heat stored in the tank during the day.

Figure 56: Non-return valve.

To avoid this phenomenon, it is necessary to install a non-return valve between the pump and the collector; this valve has to be well proportioned so as to prevent its opening by the sole thrust force of the thermal vector fluid. In this way, the storage tank will not be cooled by the collector when the pump is not running [5, 9].

Regulating power unit The regulating unit of a thermal solar system controls the running of the circulation pump to exploit the solar energy to its fullest. Often, we talk about simple electronic power units based on the temperature difference. This kind of power unit installed in standard systems (collectors on the roof and storage tanks in cellars) are provided with two temperature sensors. The first sensor is installed inside the collector, at the hottest point of the solar circuit, and the second sensor is installed inside the tank, connected with the heat exchanger of the solar circuit.

Figure 57: Electronic power unit.

The temperature values detected by the sensors are compared by a control device: the pump is run by a relay when the intervention temperature is reached. The correct definition of the intervention temperature comes from different factors. Generally, the longer the circuit pipes are, the larger is the temperature difference or the delay in the intervention. To make the pump work, standard directions suggest that the temperature difference between the solar collectors and the storage tank should be between 5°C and 8°C. Instead, the pump switches off when the temperature difference reaches 3°C. It is also possible to insert a third optional sensor which measures the temperature of the upper part of the storage tank [33].

Temperature sensors The efficiency of the pump's intervention mostly depends on the position of the thermal sensors. Each collector's sensor is positioned on the storage pipe or directly on the absorber (the part of the collector which absorbs solar radiation), near the lot's exit. However, the thermal sensor has to record the absorber temperature and communicate it to the regulating power unit, even in a stalemate situation, i.e. when the pump is not working.

The storage tank's sensor has to be placed at the same height as the heat exchanger and it can be an immersion sensor or a contact sensor. The sensor for the auxiliary heating communicates to the power unit when this intervention is necessary and has to be put at the same height as its respective heat exchanger [33].

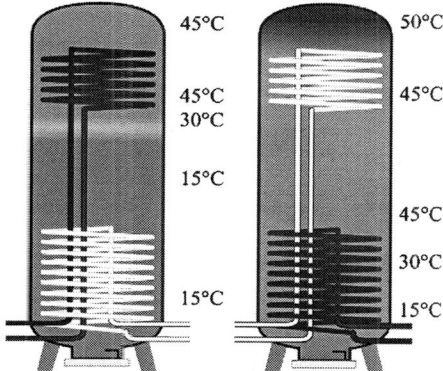

Figure 58: On the left, the processing of the auxiliary heating can be seen; on the right, the storage tank is heated by the solar circuit's exchanger.

Expansion vessel It absorbs the thermal vector fluid's expansion. The size of this component depends on the fluid quantity inside the circuit, since the vessel must be able to contain the fluid's dilatation between 4°C and 90°C.

Figure 59: Expansion vessels.

Let us assume a system with a 100-l tank and with collectors and pipes which can contain a water volume of 20 l. The expansion vessel's volume must be able to absorb the dilation of 120 l of water which occurs at the just said temperatures. The pipes which connect the expansion vessel to the system must not be insulated [5, 6, 17, 33].

Security valve Security valves protect the system when the pressure increases because of various reasons, such as superheating. Such circumstances might arise when the circulation pump is broken or is not working due to a power black-out. So the fluid's temperature inside the collectors and other circuit components may increase until the formation of steam which is then released by the security valve.

Figure 60: Security valve.

The valve must not operate during the system's normal processing and therefore it has to be set at a higher pressure than the maximum pressure of the circuit. For example, it is chosen as a pressure of 600 kPa (6 bar) if the circuit's pressure is set at 550 kPa (5.5 bar) [5, 17].

'Jolly' valve To avoid air storage inside the pipe and thereby the reduction of fluid delivery and thermal exchange, there must be vent-holes in the upper parts of the circuit.

Figure 61: 'Jolly' valves.

These vent-holes (jolly valves) can work automatically or manually. In the open circuit systems, the jolly valve should be left open since air continually enters [5, 33].

Flow regulating valves Especially for medium- and large-sized systems, these valves are inserted in every collector's row to balance the flows inside the different branches of the circuit. In this way, uniform performance from different parts of the system is guaranteed [5].

Intercepting valves The function of these valves is to interrupt the flow and insulate certain circuit elements (such as valves or pumps), when maintenance is needed or when there are security problems. They are installed at the upper and lower part of each system's element [5].

Emptying taps Manual emptying taps are installed at different circuit points. Generally, there is one in every collector's row. To allow the gradual emptying of the fluid contained inside the circuit, it is necessary to 'fix' an emptying tap between two intercepting valves [5].

Three-port valves Three-port valves allow combining two flows (mixing valves) or separating a flux into two parts (diverting valves) [5].

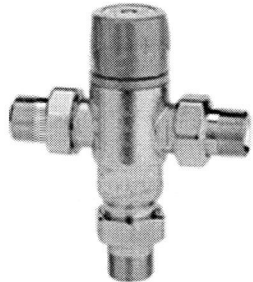

Figure 62: Three-port valve.

The pump, the non-return valve, the expansion vessel and the security valve are offered in the market as a 'pumps and security' pre-set group. The expansion vessel and the security valve have to be installed in any case to avoid interruption between them and the collector [6].

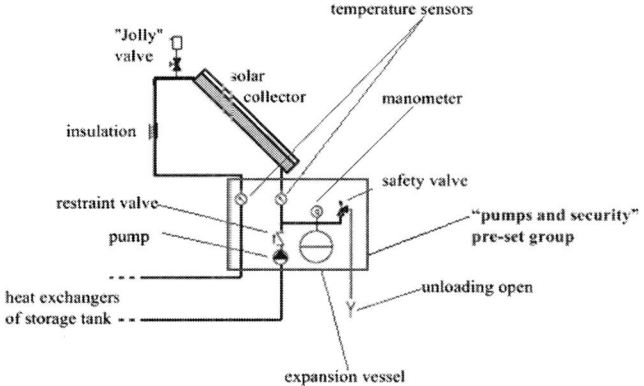

Figure 63: Pumps and security group.

2.2.4 The storage tank

As we are already aware, the solar source is aleatory and discontinuous and hence frequently production and demand are out of phase. Therefore, for the thermal solar system it is necessary to install a storage system to store the heat and make it available when it is needed. The best solution would be to store the heat produced during the summer and use it in the winter (seasonal storage), but as we have already stated in point 3 of par. 2.2.2.2.2, the technological solutions based on this principle are not convenient for limited domestic use because of its cost and logistics. Currently, the most common storage systems are the ones which allow the storage of the heat efficiently for a day or two. Storage tanks can be classified depending on their final utilization and the kind of insulation used [5, 9].

Table 4: Classification of storage tanks.

Type	Pressurized tank	Non-pressurized tank
Drinkable water storage tank	Stainless steel Enamelled steel Plastic-covered steel	
Buffer Storage tank	Steel	Plastic
Combined storage tank	Steel/Stainless steel Steel/enamelled steel	

2.2.4.1 Storage tank materials Pressurized tanks are made of stainless steel, enamelled steel or plastic-covered steel. Stainless steel tanks are lighter and last longer, but they are much more expensive than enamelled steel tanks. However, stainless steel is easily corroded by water with a high chlorine content. To reduce the corrosion risks, this kind of tank is generally provided with a magnesium anode which has to be replaced periodically. Non-porous plastic-covered steel tanks are also available in the market and they cost less than the other tanks. However, this storage tank cannot withstand temperatures higher than 80°C. Plastic-covered tanks are characterized, in fact, by a lower reliability level compared with other constructive typologies [9].

2.2.4.2 Sanitary water storage tank Figure 64 shows the storage tank installed in standard solar systems. In this storage device, there are generally two heat exchangers: the solar exchanger, which allows the thermal exchange between the thermal vector fluid inside the solar system and the fluid inside the tank, and the additional exchanger, which allows heat transfer from the integrative heating system (e.g. a central-heating boiler) to the fluid stored inside the tank.

Moreover, in the lower part of the storage tank there is a connection to the water pipes for the supply of cold water. The operating pressure inside the pressurized storage tanks is about 4–6 bar.

As regards the choice of the storage tank's volume, we generally consider 40–100 l/m^2 of flat collector surface. Concerning the proportioning of the solar system, while for the big-sized systems we refer to values that are near the lower limit of the above interval, it is vice versa for small systems. Large-sized storage

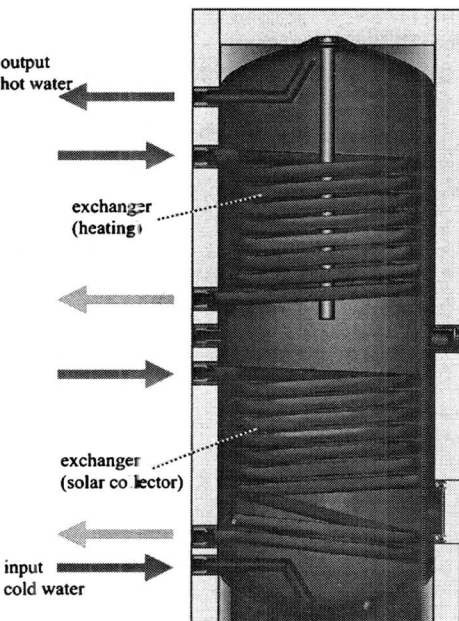

Figure 64: Storage tank installed in standard solar systems.

tanks can contain larger quantities of energy; however, this choice also involves larger heat losses and frequent starting of the integrative heating system. This happens because the heating of larger quantities of water requires more energy. As regards sanitary water storage tanks, it is important to take into consideration the calcareous encrustment problem, which may form at high temperatures in the exchanger: for this reason, in the solar systems used inside houses the tank temperature should not be above 60–70°C [9].

2.2.4.3 Storage tank shape If the storage tank works properly it should have different water layers inside. The creation of these layers is possible thanks to the variation in fluid density at different temperatures. Actually, hot water which is 'lighter' is stored in the upper part of the tank while the 'heavier' cold water is stored at the bottom of the tank. This layering effect is an essential requisite for the good processing of the solar system. As soon as the hot water is requested by users, for example, for showering, the cold water flows into the tank from the pipes and mixes with the previously heated water. To limit this undesired effect and to maintain the temperature layering for as long as possible, the storage tank (generally shaped like a cylinder) should be tall and narrow.

These conditions can be realized using vertical storage tanks whose height-diameter ratio is at least 2.5:1 (in monobloc natural circulation systems the storage

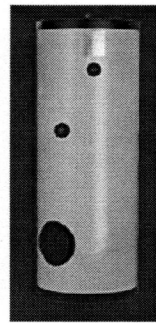

Figure 65: Example of a solar storage tank.

tank is generally horizontal because of aesthetic and encumbrance reasons). Low temperatures in the lower part of the tank guarantee a high performance of the solar system even in case of insufficient radiation and low temperature of the thermal vector fluid.

Before installing a vertical tank, it is important to make sure that its height is compatible with the place where the storage system will be placed [5, 9].

2.2.4.4 The cold water inlet device in the storage tank This particular inlet device, when suitably installed in connection with the adduction pipe, weakens the strong motion of cold water flowing into the tank and limits the risk of its mixing with the warmer water layers.

2.2.4.5 Hot water collection In traditional storage tanks, hot water is collected from the upper part of the tank; because of the layers' phenomenon and the outlet pipe being located in the upper part of the tank, we are always sure to collect the hottest water. After the collection, a part of this hot water stagnates inside the pipes getting cold. It is possible that this cold water can flow back into the upper part of the tank where it mixes with the hot water which is stored there. This causes heat dispersions of the relevant entity (see Fig. 64). To avoid this drawback, it is possible to direct the lot tube downwards making it pass inside the storage tank or outside across the insulated layer which covers the tank [9].

2.2.4.6 Heat exchangers and respective connections The solar circuit's heat exchanger should be installed in the lower part of the storage tank to ensure that the thermal exchange occurs inside the water volume present at the bottom of the tank. The heat exchanger for the integrative heating system is generally placed in the upper part of the tank, to guarantee quick heating of the water volume at a temperature (corresponding to the daily requirements) without resulting in a temperature increase in the lower part of the tank, where the solar circuit's exchanger is installed. This disposition of the exchanger ensures that the thermal exchange in the lower part of the tank, where the water is the coldest, occurs with the highest efficiency even when the solar circuit fluid does not reach the highest temperature [5, 9, 33].

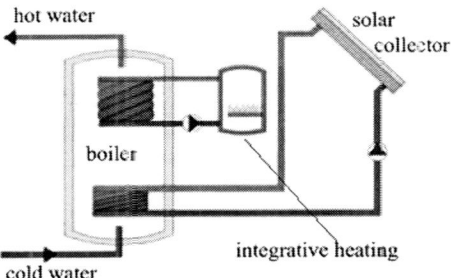

Figure 66: Working scheme of a solar system accumulator.

2.2.4.7 Storage tank insulation The purpose of the storage tank insulation is to reduce the heat dispersions to the outside environment to its minimum. To have a storage tank insulation which efficiently limits heat losses, the following characteristics are needed:

- It should be 10 cm thick on the sides and 15 cm thick in the connections with the upper surface.
- It should also cover the bottom of the tank.
- It should be perfectly adherent to tank's sides to avoid losses by convection.
- It should be made of materials which do not contain CFC and PVC and have low thermal conductivity (<0.035 W/m K).

Thermal dispersions in an insulated storage tank must be lower than 2 W/K. To limit these losses, it is very important to make sure that the thermal covering in connection with flanges and pipe fittings is hermetically sealed.

The tank linings that are currently available in the market can be flexible (expanded polyurethane foam, fibreglass, etc.), inflexible (they could be used outside for retrofit interventions) or by direct injection with a plastic or metal covering [5, 9, 33].

2.2.4.8 Few specific solutions
Buffer storage tank This kind of storage tank can be made of steel (pressurized tanks) or plastic and it is mainly used for room heating. In this case, the fluid inside the storage tank is withdrawn, heated inside the boiler and put back inside the tank; at this point, the warmed fluid is once again withdrawn from the tank and sent to the radiators. This solution is adopted to improve the boiler working conditions and so the boiler is not forced to work in the *stop-and-go* mode. Actually the boiler warms up the stand-by water volume which is kept at a certain temperature inside the storage tank and so it can remain switched off for a long period of time [9].

'Tank in tank' storage tank The combined solar systems are conjointly used for both warming up sanitary water and room heating. This system often uses a type of storage tank called tank in tank, which consists of a buffer tank inside which there is a storage tank for drinkable water. The latter tank gets the heat through its own sides from the fluid contained inside the outer tank. The storage tank with sanitary water is located in the upper part of the buffer tank where the water is maintained at a certain temperature by an integrative heating system. At the bottom of the buffer tank, instead, there is the solar heat exchanger (see Figs 67 and 68) [9].

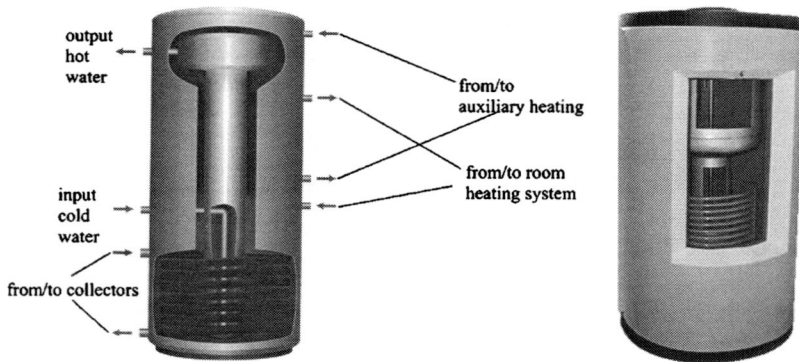

Figure 67: *Tank in tank* storage tank.

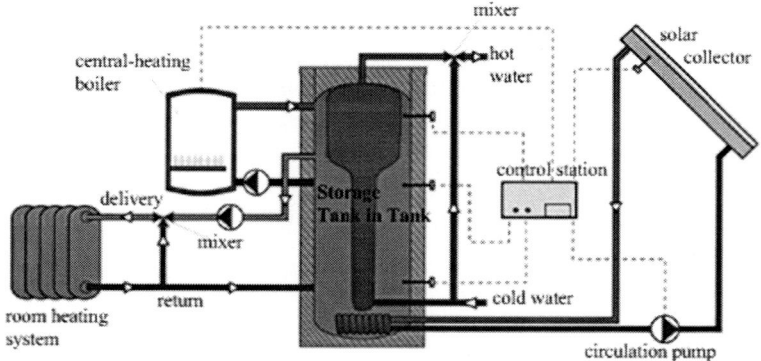

Figure 68: Scheme of a combined solar system.

2.3 Passive solar heating systems

Low-temperature thermal solar systems, which have been described until now and in which the energy transfer from the storage place to its utilization is realized by fluids moved by pumps and ventilators, are also called active systems. By the expression 'passive heating systems' we generally mean all applications where the thermal hygrometric well-being conditions are obtained only by solar energy which is used without employing any conventional heating systems requiring electricity or fuel. In merely passive systems, even the heat distribution and removal are realized by natural the phenomenon of conduction, convection or radiation, rather than using forced systems. Passive heating systems require the installation of wide glazed surfaces for solar energy interception and also structures with high thermal capacity storage function.

The efficiency of these systems is limited to the width of the glazed area which has to be correctly oriented, to the efficiency of the thermal storage realized by the walls and inner floors and eventually to the stored heat distribution towards the building parts characterized by scarce solar radiation. Currently, the realization of passive solar heating systems capable of guaranteeing the comfort conditions required in every inner room of a building seems to be an impossible goal both in cold climate zones and in mild climate zones such as Italy. However, its contribution to the reduction of the yearly heating requirement could be relevant. Passive solar heating systems can be classified on the basis of the mechanism of energy transfer towards the heated room as follows [1, 3, 4]:

- Direct gain systems
- Indirect gain systems
- Isolated gain systems

2.3.1 Direct gain systems

The direct gain system is the most common and simplest solution for a passive solar heating system: solar radiation enters the room through a glazed surface and directly warms it (Fig. 69). So, the living space works as a solar collector, but it must have the means and structures that are capable of absorbing and storing the intercepted thermal energy to keep the internal air temperature constant as much as possible. In this way, the daily overheating and the excessive decrease

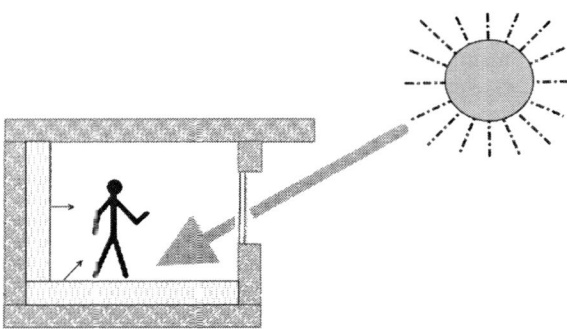

Figure 69: Direct gain system.

in night temperature can be reduced. A direct gain system needs a wide glazed surface oriented southward to allow transfer of the winter solar radiation through direct communication with the living space. The southward orientation generally allows intercepting the greatest quantity of solar energy during winter, whereas in summer since the sun is very high, the transmitted radiation is less and it can be minimized by a suitably proportioned horizontal object (overhang). The choice of window components is very important in planning the solar heating system. Windows with a high heat transmission coefficient are preferred to maximize the quantity of intercepted radiation when the radiation is very poor and also to restrict heat losses.

A wide glazed surface oriented southward can cause overheating inside the house during the day and excessive inner air temperature fluctuations when there is no direct solar radiation (during the night or when the sky is cloudy). To solve these problems, it is important to use a thermal mass that is connected with the walls and floors whose surfaces and thermal capacities are well proportioned and also well positioned to intercept solar radiation and store thermal energy. During the day, the heat produced by the intercepted radiation is not completely released into the room; it is partially stored and released later after a delay of a few hours to stabilize the air temperature of the house. The storage thermal mass is generally made of masonry. Masonry materials for thermal storage are characterized by a high thermal capacity: cement blocks, concrete, bricks, stone, etc. To restrict dispersion of stored energy inside the thermal mass, the brickwork walls are insulated on the outside, while floors are realized with a perimeter or an extrados insulation [1, 3, 4].

2.3.2 Indirect gain systems

As regards indirect gain systems, solar radiation does not directly enter the room that has to be heated up, but it falls on a thermal mass which is placed between the Sun and the living space. The solar energy absorbed by that mass is converted into thermal energy and then distributed inside the room in different ways. By the position of the thermal mass, we can distinguish two kinds of indirect systems: solar walls where the thermal mass is contained inside a wall and roof-ponds where the thermal mass is put on the roof of the room which is to be heated up. Indirect gain systems need a wide glazed surface oriented southward and the thermal mass used for storing the absorbed energy is placed behind it at a distance of at least 10 cm.

The storage is normally made of brickwork or water (with the latter put inside metallic, plastic or concrete waterproof containers) [1, 3, 4].

2.3.2.1 Brickwork solar walls Solar radiation received by a solar wall which is painted with a dark colour is absorbed causing a superficial heating up (Fig. 70). This heat, which is like a temperature damped wave, is transferred by conduction to the wall's inner surface and from there it spreads all over the room by radiation and convection. The delay and the temperature wave damping depend on the storage material and thickness. Dispersion of stored heat towards the outside is resticted by the insulation created by the air space between the glazed surface and the solar wall.

Trombe's wall (Fig. 71) is different from the solar wall owing to the presence of air holes in both the lower and the upper parts of the wall. In this way, the activation of a mechanism for natural circulation of air through the heated area is favoured. The warm air volume between the glazed surface and the thermal mass can reach high temperatures (about 65°C). The air holes in the upper part of the storage mass allow warm air to move up and enter the room, while the colder air which is inside the room is recalled inside the collector through the holes in the lower part of the storage mass. The openings should be located by means of dampers to prevent the reverse movement during the night [1, 3, 4].

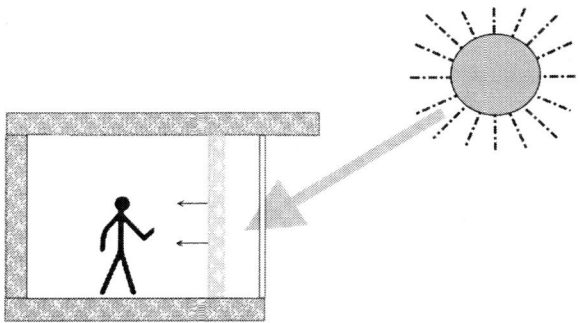

Figure 70: Indirect gain system: solar wall.

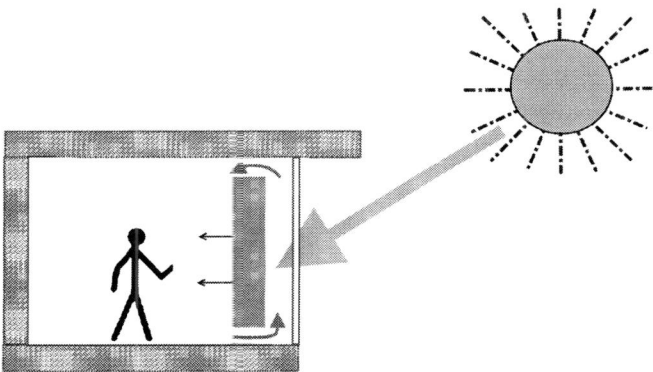

Figure 71: Indirect gain system: Trombe's wall.

2.3.2.2 Water wall The processing by a water wall (Fig. 72) is based on the same principle which regulates the processing by a solar wall, excepting that heat transmission through the wall also depends on thermal convection and not only on conduction. Because of the high thermal capacity of water and the inner convective currents, which make it an almost isothermal heat accumulation, the system can work with a higher efficiency compared with brickwork solar walls. One of the most important problems is where to confine the liquid. Until now, bottles, tubes, watertight tanks, barrels, drums and cement walls filled with water have been used as containers (see Fig. 72) [1, 3, 4].

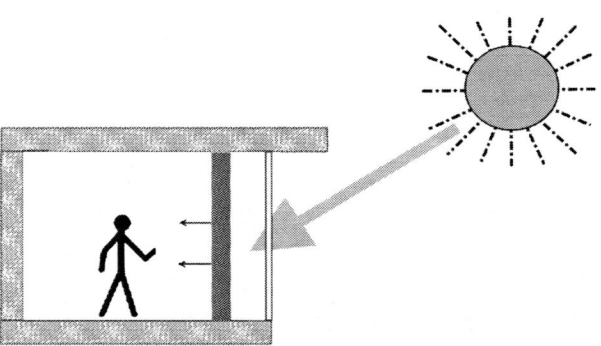

Figure 72: Indirect gain system: water wall.

2.3.2.3 Roof-pond As regards roof-pond passive systems, the thermal mass is placed horizontally on the building's roof (Fig. 73). The storage medium is water which is enclosed in small bags similar to little mattresses. They completely or partially cover the roof which works as the ceiling of the rooms that are to be heated up. Water containers have to be placed in direct contact with the ceiling which sustains them to make the heat exchange between the inner room and the storage easier. During the hot season, the storage is exposed to solar radiation during the day; the intercepted energy is then transferred by conduction through the roof structure and directly exchanged by radiation from the ceiling of the room to be heated. During the night or during cloudy days, a mobile insulation mechanism covers the hot water and restricts its heat dispersion. In contrast to the systems with solar walls, systems with water walls are not always provided with a transparent cover to put on water. The use of translucent containers or glazed surfaces put on the water mirror is an efficient solution to reduce sensitive and latent (evaporation) heat losses when climate is very cold.

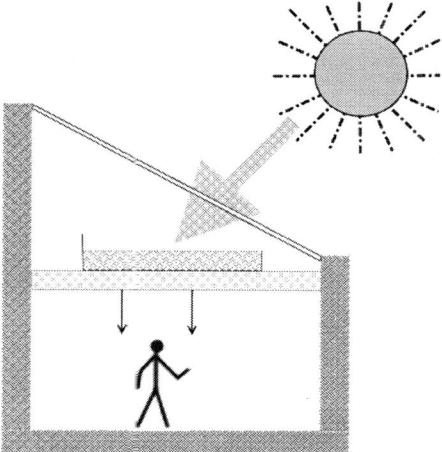

Figure 73: Indirect gain system: water wall (*roof-pond*).

In places where there are high thermal ranges between day and night and where humidity is very low, the water storage on the roof can also be also for summer refreshing by insulating it during the day and exposing it during the night. The utilization of the water wall also presents numerous problems: besides the extra structural costs, the system does not guarantee sufficient advantages at high latitudes because of the reduced solar radiation intercepted by the horizontal plane; moreover, the stored heat can be spread by radiation only over the floor below the roof [1, 3, 4].

2.3.3 Solar greenhouse

A solar greenhouse, which is set against a building or is made out of a building, consists of a closed glazed space located on the south side of a house which is separated from it by a thermal accumulator wall (Fig. 74). The greenhouse can be used as both a direct gain con-warmed space and an indirect system since the rooms next to it receive heat through the intermediate wall which works as a storage. It is also possible that the rooms receive heat from the air in the greenhouse through a natural or forced ventilation system. Solar greenhouse planning can follow different criteria: if it is considered as a cheap extension of the house where people live for the greatest part of the year, it will be necessary to employ a big storage mass placed both on the walls and also on the floor and some movable insulation panels for the night. Instead, if the greenhouse is seen as a solar wall system, with an air space which is a few metres wide rather than a few centimetres, it should be planned to ensure that the greatest quantity of intercepted energy will be taken from the air space to heat up the adjacent rooms. In this case, a forced air change using the greenhouse to pre-heat the incoming air could be also planned (Fig. 75).

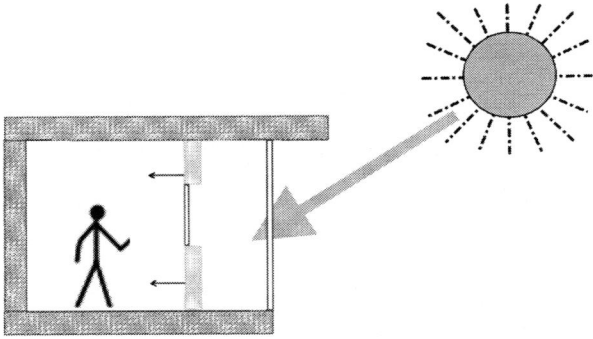

Figure 74: Solar greenhouse.

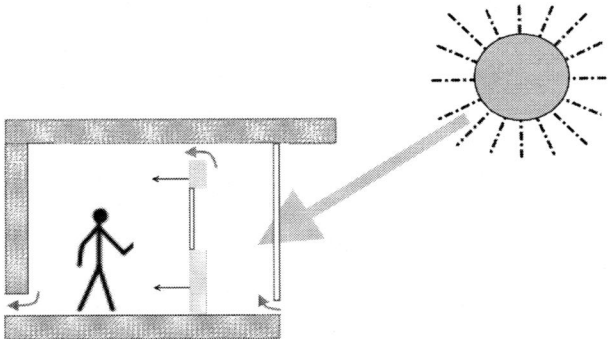

Figure 75: Greenhouse with a controlled ventilation system.

Solar greenhouses can be realized in a wide range of geometrical configurations. It can be considered as a simple addition to a wall, as a semi-jutting out element or as an element which is set in a building (with three of its sides surrounded by living spaces). Moreover, the solar greenhouse can be considered as a structure which covers the entire width of the house and is a single storey or two storeys. Even a greenhouse which is isolated from the building structure can supply thermal energy to the building through a system of ventilators and grooves.

Eventually, a correct solar greenhouse plann has to restrict the inner overheating phenomenon during the summer to its minimum. The simplest technique is one which allows ventilation directly from outside by opening the glazed windows, but the use of screenings or glazed surfaces fitted with sun block control is also possible [1, 3, 4].

2.3.4 Isolated gain systems

The third and last approach to passive heating is the isolated gain system. In this system, the solar collector and the storage are thermally insulated from the rooms that are to be heated up. The system can run apart from the building which draws energy only when heat is required. Where systems are completely passive, the energy transfer from the collector to the room or to the storage and from the storage to the room happens only by natural processes and not by forced processes such as convection and radiation. The most common technique is the one to create natural circulation systems composed of a flat plane collector and a thermal accumulator tank. The thermal vector fluid is normally air. An air radiator system (Fig. 76) uses a glazed collector located in the most suitable position to get the greatest quantity of the Sun's radiation, but it must be distant and below the thermal storage tank.

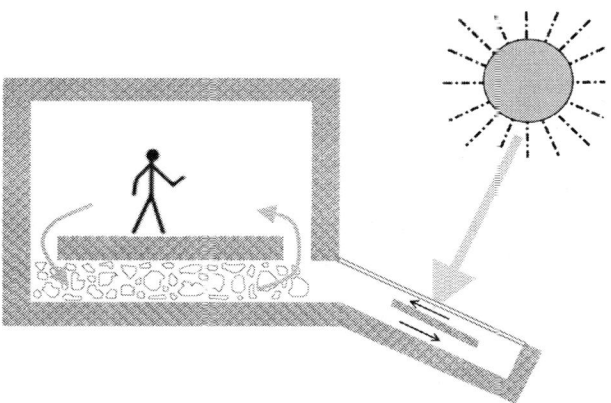

Figure 76: Isolated gain system: radiator system.

The absorbed heat warms up the air which, because of the density gradient, moves up and enters the storage (made of either a compact conglomerate mass or an incoherent bed of stone) thereby heating it up. The stored heat is then distributed over the air in the room by convection. The thermal storage mass can be put below the floor of the building, below the windows or inside pre-fabricated plugging elements. The space orientation of the building is less important for the system's efficiency compared with the other kinds of solar gain [1, 3, 4].

3 Medium-temperature solar thermal technology

Medium-temperature technology includes systems which are able to reach temperatures between 100°C and 250°C. The most common application of the medium-temperature solar thermal system is represented by the solar oven (see Fig. 77): a parabolic reflector (composed of aluminium sheets mounted on a zinc-plated steel structure) concentrates the solar radiation towards a single point which works as a cooking-stove. At this point, a pot is placed which warms itself and cooks the food contained inside. Using a solar oven it is possible to reach the same temperature as a traditional cooking-stove (about 200°C).

Figure 77: Solar oven.

A solar oven with a diameter of 1 m takes nearly 18 minutes to boil 1 L of water, while it takes only 9 minutes if the diameter is wider (1.4 m). The reflector can be oriented on the basis of the Sun's position so that it is possible to cook from morning to afternoon and even to exploit the shortest moment of radiation. In Italy, the use of solar ovens is not common; they represent a very small slice of the market and their use is restricted to the those who consider it a hobby. In countries where lack of energy resources is a daily problem (such as Africa), the solar ovens can have good applications [12, 13, 18, 41].

In spite of the various advantages offered by thermal solar systems, their great potential has not been exploited much in the industrial sector. Thermal solar systems can partially meet the heat demand for low- and medium-temperature (up to 250°C) processes, which are typical of a few industrial sectors such as the chemical, food and textile industry. The thermal solar collectors that are now available now in the market, which we analysed when we talked about low-temperature solar thermal systems (par. 2.2), can reach temperatures of 100°C. As regards applications which need higher temperature (up to 250°C), the experiences are limited and suitable collectors do not exist. In 2003, the International Energy Agency (IEA) started a research project called *Task 33/IV* which aims to find more promising industrial applications in the thermal solar field and also to calculate the overall potential of thermal solar applications for the production of medium-temperature process heat. One of the Task 33/IV activities is research, developed together with the industry, on new collectors which can produce processed heat between 100°C and 250°C (a temperature range that is consistent with several industrial processes) [42–44].

At present, the collector typologies which are more promising in the medium-temperature field are:

- high efficiency glazed flat plate collectors: these are flat collectors with double antireflection glass;
- linear parabolic collectors, similar to the ones used in the high temperature field but much smaller (these will be analysed in par. 4.4.1).

- static concentration solar collectors: these are flat plate collectors or more frequently evacuated tube collectors characterized by static mirrors (fixed) for the concentration of solar radiation.

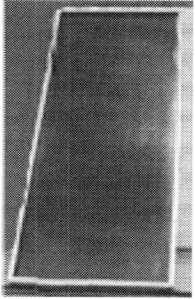

Figure 78: Double-glazed flat plate collector with antireflection glass.

Figure 79: Linear parabolic collector.

Figure 80: Static concentration collector.

4 High-temperature solar thermal technology

4.1 Concentratiing solar power technology: clean energy for power tenability

Energy availability has always been an essential component of human civilization. In the last 150 years, the yearly average rate of world energy consumption has grown by about 2.3%. The energy requirement of human beings, mostly satisfied by fossil fuel, has grown so much that it has overcome sum of the thermal energy coming from the Earth's core and from the ties induced by the Sun and the Moon. The endogenous energy of the Earth has been more than doubled by human activities. However, it is important to underline that the overall human energy consumption is only 1/10,000 of the energy received on the Earth from the Sun. Solar radiation, despite its scarce density, remains the most rich and clean energy source on the terrestrial surface [45–47].

This statement, together with the exhaustion of fossil fuels and the growing environmental risks, leads us to seriously consider solar energy as one of the most important candidates for the planet's energy tenability project.

The greatest part of the solar source's potential can be found in the so-called 'sun belt', which is the area of the planet that receives the highest quantity of solar radiation, as shown in Fig. 81. In particular, Northern Africa and the Middle East have large areas with a high level of solar radiation which is suitable for the installation of a large number of solar thermal systems as they cannot be used in any other way. These countries spontaneously stand out as candidates for the intensive development of solar energy [45].

From the recent MED-CSP research, commissioned by the German Ministry of Environment Policies and carried out at the Aerospatiale Centre DLR, it has

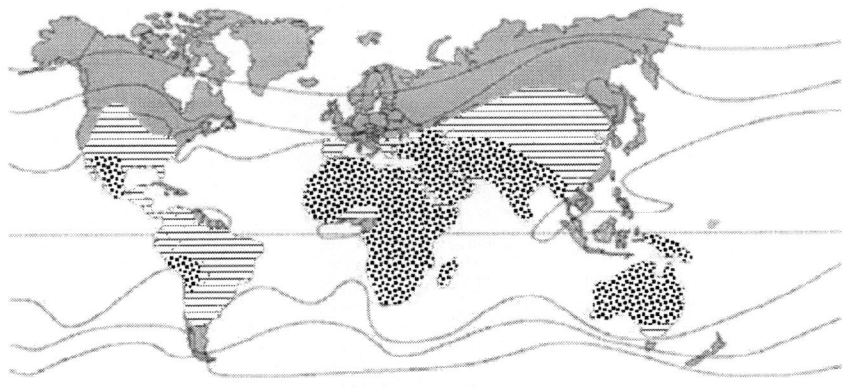

Suitability for the implementation of solar concentration
excellent ≡ good ∎ unfit

Figure 81: World map of solar radiation underlining the *sun belt*.

been shown that the solar energy potential available in the countries on the Mediterranean coast is much larger than the actual electrical energy consumption of the area which includes Southern Europe, the Middle East and Northern Africa. Besides the DLR and other German organizations, other centres of research such as NERC (Jordan), CNRST (Morocco), NREA (Egypt) and NEAL (Algeria) contributed to the development of this study. This project shows that a common interest could link European countries (energy importers and technology exporters) with the Northern Africa countries (owners of fossil fuel resources and solar energy who will see a significant growth in their energy consumption in the next few years) [45, 48]. To exploit this huge solar energy potential, the concentrating solar power (CSP) technology is very useful. This technology exploits solar radiation, concentrating it using mirrors for electricity production and also for the realization of high-temperature chemical processes, for example, the production of hydrogen. A more detailed description of CSP technology is given in par. 4.4. This technology can be considered as a competitor of the photovoltaic technology, which is already common and is growing in Europe (actually, this is only partially true). On this matter, two aspects have to be considered: first, the photovoltaic technology exploits both the direct radiation and the spread (diffuse) radiation and so it is also suitable in areas such as Northern Europe where direct solar radiation is scarce; second, it is fit for many different applications (from a few watts of a mobile phone's solar battery charger, to the megawatts of dedicated solar systems, passing through a few or tens of kilowatts used in many applications in the residential or civil field). CSP technology, on the other hand, exploits only the direct radiation and lends itself poorly, excepting in a particular situation or only in thermal applications, to the realization of small dimension systems. As regards systems that have a capacity of about or more than a megawatt and that are used in areas with strong direct radiation, CSP technology allows achieving a cost for electrical energy production which is much less than that for the energy produced using photovoltaic technology, and this advantage, in this specific case, is destined to last for a long time unless there is a radical technological improvement in the photovoltaic field [45, 48].

Considering only the European part of the Mediterranean area, we can observe a sort of integration between the two technologies: the photovoltaic technology is installed where there is lesser direct radiation and in applications which require from a few kilowatts to a hundred kilowatts of power. On the other hand, CSP technology is installed in areas with stronger direct radiation and in medium–large power systems (starting from a megawatts of power). It is also possible to think of a situation where Europe increases its own consumption of 'green electricity' by taking it both from the different renewal sources available in the area and from the solar energy imported from the most suitable regions.

To highlight the significance of this project, which has got many prospects, it is necessary to point out that the density of the solar energy received in the Mediterranean southern coast and its territorial characteristics allow to cut the solar energy production costs, which are then invested to produce solar energy in the southern

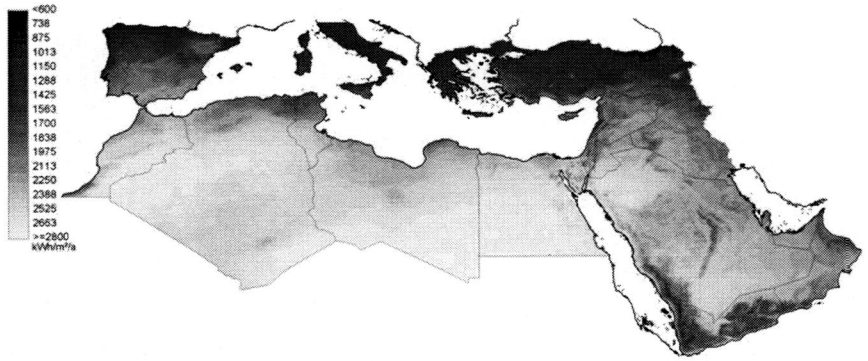

Figure 82: Map of the solar direct radiation in the Mediterranean area: lighter spots indicate regions with the strongest radiation.

European area; moreover, large areas which cannot in any case be used for agricultural purposes are available on the coasts of Northern Africa and the Middle Eastt. Since the cost for high-tension and continuous line electricity transmission for a distance of about 1000 km, of which 100 km is by submarine cable, is about 0.7–1.5 c$/kW h, it is not illogical to think of realizing in those regions – inside a wider project of social and economic integration – solar energy production capable of satisfying the growing requirement for electricity in Northern Africa and also a part of Europe [45–47].

With the regard to this matter it is important to underline how, despite mere energy considerations, the so-called 'Mediterranean electric belt' has been studied for a year, which will shortly allow the complete electrical interconnection between the Mediterranean countries and the European electric net.

Considering what we have just stated, it is clear that the wide exploitation of the solar source in the Mediterranean area is a very important topic with a high political and economic agenda since it has important consequences in terms of bringing about the integration of the northern and southern parts of the world and also in terms of development of pacific relationships.

As regards future prospects, the direct production of hydrogen, mainly by CSP technology, will allow the sun belt to increase the production of energy. Presently, a few Southern European areas, Spain in particular and also southern Italy, have a fairly good potential, which allows them to exploit the CSP technologies to increase the renewal quantity of electricity production.

Spain in particular is in a favourable situation because of both the presence of a huge energy potential and the remarkable experience it has with experimental activities developed since 1981 at the *Plataforma Solar De Almeira*.

As regards Italy, currently, there are no accurate studies on the energy potential which can be exploited using the CSP technology. The MED-CSP study [48] estimates that the 'technical exploitable' potential is about 88 TW h/year while the 'economically exploitable' potential is about 5TW h/year; actually, these figures refer to very approximate values. Despite this, it is clear that the principal aim of using the CSP technology is to obtain advantages, in addition to economical advantages, by the exploitation of the energy potential in the areas that are rich in the solar source. This is more justified if we consider the case of Germany which has been pursuing the development of this technology for many years although its economically exploitable energy potential is almost non-existent. Since the primary source is free, it is also important to underline that the total turnover related to energy production from solar energy benefits those who realize this and take care of the production systems; those who own the know-how are then designated to exploit the biggest part of the business connected with solar energy production [45].

4.2 Prospects of CSP technologies

The predicted development of CSP applications follows on after 20 years of the development of the Aeolian applications; a possible trend could be the world achieving 5000 MW by 2015. The overall 'portfolio' of the CSP systems planned at different levels in the world accounts for about 1562 MW; adding to it both the predicted 28 MW coming from the Italian Archimedes Project (described in par. 4.5.3) and the portfolio of the Global Environment Facility (GEF, an independent financial organization connected with the World Bank and also with the Environment Project of the United Nations; it was founded in 1991 to help developing countries with projects and programmes aimed at protecting the world environment) projects which are currently foreseen will add another 130 MW_e, which results in a potential world portfolio for the short to medium term of over 1700 MW.

Of this 1700 MW, 300 MW is considered to be surely realized.

The factors which have resulted in a reduction of the levelled electrical energy costs produced by these systems, valued by the GEF, are shown in Fig. 83; it has been forecast that the levelled energy cost (LEC) will be reduced from the current 16 c$/kW h to about 6 c$/kW h by the year 2025, also reaching by that date the predicted cost for the fossil fuel systems. Other organizations have predicted even lower costs (till 3.5 c$/kW h).

The fulfilment of these developmental forecasts will mostly depend on the political and economic situation in the next few years. However, it is clear that the knowledge and diffusion of CSP technology is currently at the same stage as that of the Aeolian ones during the mid-1980s; at that time, nobody would have bet on the Aeolian energy; instead, we have currently reached only in Europe an installed Aeolian power of 34,000 MW [45, 49].

4.3 The Italian position and interest in CSP technologies

From what we have discussed until now, it should be clear that the Italian position and interest in CSP technologies are as follows [45]:

- Pursuing the technical and industrial development of this technology to guarantee green energy flow at a profitable price in terms of Euro-Mediterranean integration and globalization of environmental problems.

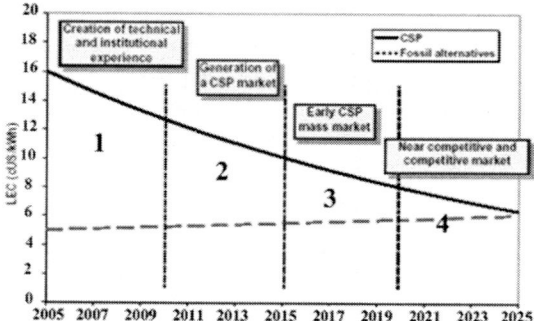

Figure 83: Trend of the predicted costs between the years 2020 and 2025.

- To ensure in time a substantial part of the future global turnover related to green energy production in terms of R&S, that is, by the creation of new Italian enterprises or by getting industrial orders for Italian firms.

4.4 CSP technology

Thermal solar power systems use solar radiation to produce heat in place of traditional fossil fuels. To get heat at a temperature higher than 250°C, it is necessary to concentrate the solar radiation. To achieve this concentration, an appropriate optical system (the concentrator) is used. This device gathers and delivers direct solar radiation to another device (the receiver) where it is transformed into high temperature heat [50]. The heat produced in this way can then be applied to different industrial processes (such as the desalination of seawater and production of hydrogen using thermal chemical processes) or to electrical energy production.

Currently, electrical energy production is the main purpose for which CSP systems are used. In this case, solar heat is used in traditional thermodynamic cycles such as the Rankine, Stirling and Brayton cycles. Until now, systems that are able to convert about 30% of the solar radiation received on the Earth's soil into electrical energy have been used. The range of power obtainable varies from 10 kW to a few hundreds of megawatts, including more than one modular system [50, 51].

As regards the concentration, solar systems can apply to different technologies; however, it is possible to point out the following processes in every one of these technologies [50]:

- gathering and concentration of solar radiation;
- conversion of solar radiation into thermal energy;
- transport and possible storing of thermal energy;
- thermal energy utilization.

The gathering and concentration of radiation, whose power density is very low by nature, is one of the principal problems of solar systems. As already stated, this process occurs thanks to a concentrator which is composed of appropriately shaped panels with reflecting surfaces. During the day, the concentrator follows the Sun so as to gather the direct component of its radiation and concentrates it inside the receiver. This latter device transforms solar energy into thermal energy, which is then given to a fluid that flows inside (*thermal vector fluid*). As we will observe from the analysis of tower systems with a central receiver (par. 4.4.2), the use of a melted salts mixture as the thermal vector fluid lets the system have a system of thermal energy accumulation before its utilization in the production process. This storing is realized by collecting the thermal vector fluid that goes out from the receiver in appropriately insulated storage tanks. In this way, solar energy, which is very variable in nature, can become a thermal energy source that is always available to users wherever required [45, 50].

On the basis of geometry and the concentrator's position, we can have three kinds of CSP systems [12, 45, 50–53]:

- linear parabolic collector systems;
- tower systems with a central receiver;
- parabolic dish collector systems.

4.4.1 Linear parabolic collector systems

Currently, the most suitable technology for electrical energy production by thermodynamic systems is one which uses linear parabolic collectors [45, 50, 53].

These collectors are composed of a linear concentrator with a parabolic profile whose reflecting surface follows the Sun rotating on a single axis. The concentrator is fixed on a support structure which guarantees the correct processing during windy conditions and the action of other atmospheric agents.

The reflecting panel is normally composed of a common glass mirror with an appropriate thickness. Solar radiation is focused towards a receiving tube that is placed along the parabolic concentrator's fire. The energy absorbed by the receiving tube is then transferred to a processing fluid (thermal vector fluid) generally made of synthetic oil which is drawn up inside. The heat gathered is normally used as shown in the Fig. 86 (i.e. for the electrical energy production).

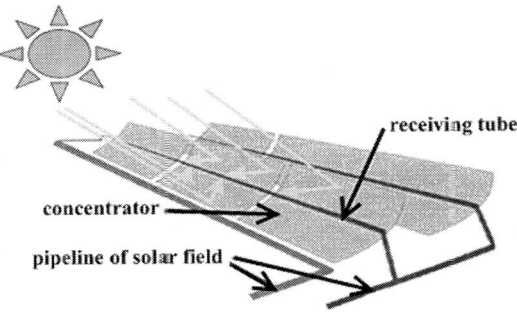

Figure 84: Linear parabolic collector systems.

Figure 85: Solar field.

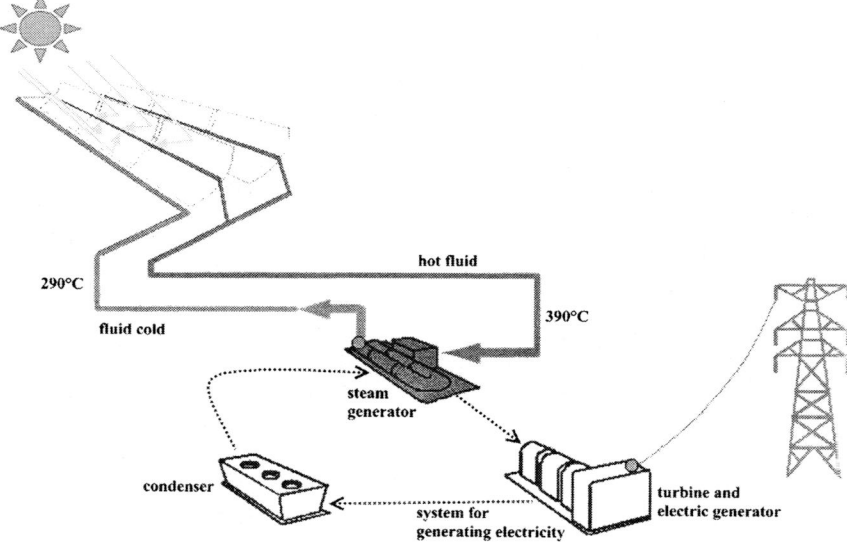

Figure 86: Schema of a thermal electrical system with linear parabolic collectors.

Figure 86 shows the processing scheme of a solar thermal electrical system with linear parabolic collectors using synthetic oil as the thermal vector fluid. In such systems, parabolic collectors are connected in series, generally in two parallel rows which are a few hundred metres long and form a string that represents the unitary module of the system. The strings as a whole form the so-called solar field (Fig. 85). The synthetic oil pumped towards the receiving tubes comes out from the warmed up solar field at a temperature of about 390°C and then feeds a power unit (which is right at the centre of the solar field): the thermal vector fluid transfers the heat to a steam generator to start the processing of an electrical turbo-generator group. After delivering the heat, the oil (at 290°C) comes back to the solar field to be warmed up once again. With the linear parabolic collector technology, a maximum system processing temperature of 600°C can be reached (but it also depends on the kind of thermal vector fluid used and on its temperature when coming out from the solar field).

Nevertheless, the use of synthetic oil as the thermal vector fluid, which is the case in almost all solar systems with linear parabolic collectors, does not allow reaching temperatures higher than 390°C (as seen in Fig. 86) which has a negative influence on the thermodynamic performance of the steam generator group. In these systems, the conversion efficiency for the conversion of solar energy directly into electrical energy is 15%. At present, solar thermal electrical systems with linear parabolic collectors have typical dimensions; the capacity of these systmes can be in the range 30–80 MW$_e$ and they can also burn a certain quantity of fossil fuel (natural gas) to produce energy when there is a lack of solar energy, so these type of systems are hybrid systems, i.e. solar-fossil fuel systems [45, 50, 51, 53].

The maturity of this technology can be proved using the example of the Kramer Junction in Mojave Desert (California), where in 1984 this kind of solar thermal electrical system (SEGS I, *Solar Electric Generating Systems*) with a capacity of 14 MW$_e$ was realized. This system uses both linear parabolic collectors and natural gas as the fuel for overheating and sustaining the system in case of low radiation or breakdown.

Figure 87: *Kramer Junction's* SEGS system in California.

An additional eight systems were constructed from 1984 to 1991, SEGS II–SEGS IX, reaching a total power of 354 MW$_e$. In these systems, the technology had been improved and costs were reduced such that the cost of electricity generated was reduced from about 30 c$/kW h (in the first system) to 8 c$/kW h (in the last system). These systems have all been producing electricity and have added more than 13 TW h (billions of kW h) to the electrical net [45, 50, 51, 53].

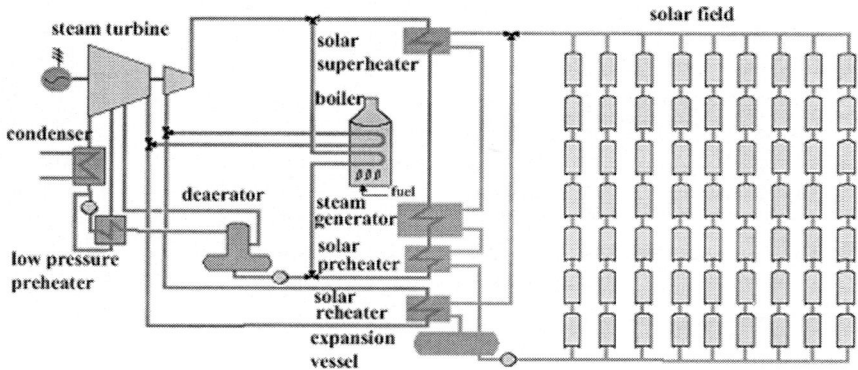

Figure 88: SEGS IX scheme.

Linear parabolic collector systems have shown a few limitations which have not allowed their wider application. The principal problems are [53]:

- electrical energy production depends on the intermittence and variability of the solar source, which necessitates the use of fossil fuels to integrate the thermal energy production and therefore the need for solar-fossil fuel hybrid systems;
- the low conversion efficiency of the systems, which is due to the limited efficiency of solar energy gathering and the low processing temperature of the thermal vector fluid (<400°C);
- the high cost of the electrical energy produced, which is a consequence of the low efficiency of the systems and the high construction cost;
- the dangers posed by the use of the working fluid (synthetic oil) which is toxic and highly inflammable at the processing temperature.

Technological development projects have been planned in many countries. In 2004, the construction cost for a solar thermal electrical system with linear parabolic collectors was about 2500–3500 €/kW$_e$, with a predicted 30% reduction in the medium term [51].

4.4.2 Tower system with a central receiver

This technology has overcome the demonstrative phase as an industrial prototype, but it has not reached the phase of trade maturity yet. The central tower system (see Fig. 89) makes use of flat reflecting panels (which as a whole form the solar field) called heliostats (Fig. 90). These panels track the Sun by rotating on two axes and concentrate the sunlight towards a sole receiver. The receiver is installed at the top of a tower (which is at the centre of the system) and inside a fluid (thermal vector fluid) flows to absorb the solar heat.

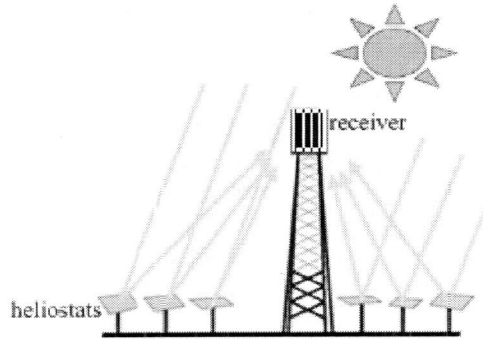

Figure 89: Solar tower system.

Figure 90: Heliostats.

The thermal energy which is made available by this process can be used in different processes, especially for the production of electrical energy.

In this kind of system, the thermal vector fluid can reach high processing temperatures (>500°C), which allows achieving high efficiencies in the conversion of solar energy into electrical energy. Generally, the transformation happens by exploiting the heat in a traditional water–steam thermodynamic cycle (see Fig. 91).

The central tower technology has shown its technological practicability in the production of electrical energy by the realization and the running of numerous small-sized systems (0 5–10 MW_e) in different countries all over the world (Spain,

Italy, Japan, France and USA). From this experience, which has come with maturity, it has been seen that the best size for these systems is in the range 50–200 MW$_e$. Different kinds of thermal vector fluid (such as water, air, melted salts) have been experimented with for years; however, until now the most suitable fluid for this technology has been the melted salts mixture composed of 60% sodium nitrate (NaNO$_3$) and 40% potassium nitrate (KNO$_3$).

Compared to synthetic oil, which is used as the thermal vector fluid in most solar thermal electric systems with linear parabolic collectors, the melted salts mixture, used in tower systems with a central receiver, has two important advantages: the fluid can reach a higher processing temperature (565°C) and it is possible to install a thermal energy accumulation system, which can be created by piping the mixture heated up in the receiver towards an appropriately insulated storage tank (see Fig. 91). The sodium and potassium nitrate mixture can be heated until a maximum temperature of 565°C is reached (when the temperature is higher than 565°C, nitrates decompose into nitrites causing potential corrosion problems), which is much higher than the temperature of 390°C allowed by synthetic oil; this higher temperature allows achieving a better performance in the thermodynamic cycle for the production of electrical energy as well.

The elevated cost, the environmental risks and the inflammability which characterize synthetic oil do not allow the storage of this warm liquid in such a volume needed to achieve an efficient thermal accumulation (actually, there is no thermal accumulation in the solar systems with linear parabolic collectors). Instead, the cheapness, non-toxicity and low environmental risks typical of the melted salts mixture make this fluid the most suitable for use in a thermal energy accumulation system, which solves the problem of the solar source variability and allows the production of electrical energy on demand so as to make the system more flexible [45, 50, 51, 53].

Figure 91 shows the processing scheme for a tower solar thermal electric system with a central receiver; it uses the melted salts mixture (described above) as the thermal vector fluid. Heliostats concentrate the sunlight towards the receiver

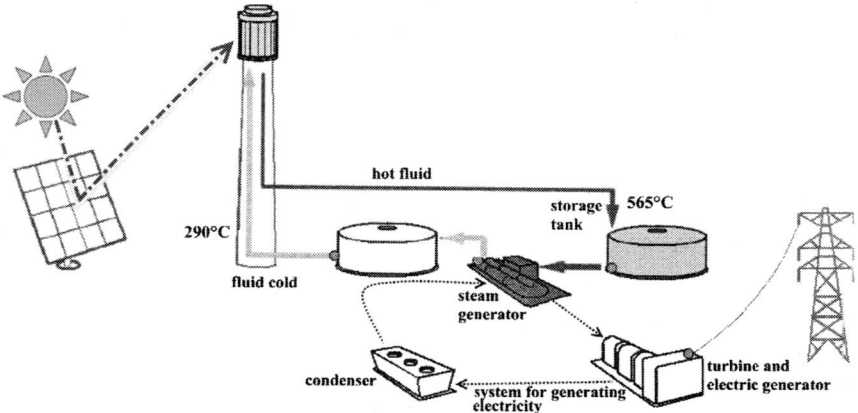

Figure 91: Tower solar thermal electric system with a central receiver.

inside which the melted salts mixture flows; this mixture absorbs the heat and reaches a temperature of 565°C. The warmed thermal vector fluid is then directed to an insulated storage tank where thermal energy accumulation takes place before being picked up for the production of electrical energy. When it comes out from the warm storage tank (565°C), the melted salts mixture delivers heat to the steam generator which feeds an electrical turbo-generator. After delivering the heat, the thermal vector fluid cools down (290°C); at this point, it is accumulated inside another storage tank, waiting to be once again directed to the receiver. Eventually, the most important improvements introduced by this kind of system compared to those brought about by the system with linear parabolic collectors are as follows [53]:

- The thermal vector fluid is more safe since sodium and potassium nitrate (well-known compost) is neither flammable nor toxic (compared to the synthetic oil).
- The increase in the processing temperature of the thermal vector fluid from 390 to 565°C improves the performance of the thermodynamic cycle.
- The introduction of thermal accumulation solves the problem of the daily variation in solar intensity. This provides clear advantages as regards the processing continuity of the turbine-alternator group and avoids resorting to the fossil fuel integration. Therefore, these systems are not hybrids but exclusively feed by renewal sources.
- The mirrors are made of compound materials (*honeycomb*) which are lighter, stronger and cheaper than the glass used in SEGS systems.

Although tower systems with energy accumulation are more efficient as regards conversion and require cheaper initial investment than the systems with linear parabolic collectors, a few disadvantages make their installation difficult on a wide scale and for high power requirements [53]:

- the conspicuous dimensions of the central tower whose height is proportional to the mirrors' field extension and the system's power;
- It is very difficult to concentrate solar radiation towards the receiver which is installed at a height of 100 m. By contrast, the focal length of systems with linear parabolic collectors is lesser than 2 m.

Without any doubt, one of the most important examples of this technology is represented by the experimental American system *Solar Two* of 10 MW$_e$ power, which was in operation from 1996 to 1999 in Dagget, California. *Solar Two* was the first system to use a melted salts mixture composed of 60% sodium nitrate and 40% potassium nitrate as the thermal vector fluid [8, 15, 42].

In Italy, as regards high-temperature solar thermal systems, the most relevant example was in the beginning of the 1980s with the construction of the world's biggest solar power plant in Adrano, Sicily. This power plant, called Eurelios

Figure 92: Impianto *Solar Two*.

(constructed within a CEE research project and thanks to the investment from an Italian-French-German society), has not been in use since an experimental phase which lasted for 6 years, from 1981 to 1987. Eurelios was able to produce a power of only 1 MW [2, 13].

Technological development projects are currently being implemented in USA, Spain (with the collaboration of a few countries), South Africa and Israel. In 2004, the construction cost for a solar thermal electrical system was about 4500 €/kW$_e$, with a predicted reduction to about 2000 €/k in the medium term [51].

Tower systems with a central receiver have and will continue to have great importance in the field of both continental and in world energy in the near future. As proof of this statement is the PS10 system with its capacity of 11 MW power. This system, which began production in January 2007, located in *Sanlucar La Mayor* (Andalusia) is the biggest European solar thermal power plant. The PS10 Spanish power plant, whose overall cost is 35 million Euro, occupies 60 hectares of land and based on predictions it will produce 23 GW h/year at a cost of 0.1 €/kW h produced. An interesting result when we consider that the best photovoltaic systems currently produce energy at a cost no cheaper than 0.23 €/kW h, which is close to cost of energy produced by fossil fuel systems (0.06 €/kW h) [25].

Observations on the thermal accumulation The introduction of a thermal accumulation system allows the elimination of the short transitory effects due to the irregularity of the solar radiation and also allows the release of the production diagram from the solar radiation diagram, as seen in Fig. 93 (where it is assumed that the electrical power installed is the same in both cases). The presence of the accumulation allows the use of a wider solar field, even if the electrical power is the same, to produce more energy and also a greater number of 'equivalent annual hours' of operation. These can go from 1500 h, typical of a system without the accumulation, to 2000–4000 h or more in a system with accumulation. A very big storage tank would virtually allow the continuous production of energy. Actually, it is more appropriate to limit the accumulation to a storage tank which would allow 5–10 h of nominal power processing. This will allow users to plan the production of electrical energy to its best, concentrating it during periods of high requirement (also increasing its trade value). In fact, as can be seen from the diagram in Fig. 94, the requirement of electrical energy in Italy has its peak during the evening–night hours and so it is delayed by nearly six hours from the solar radiation peak. This aspect is often more evident in developing countries. Generally, based on the weather forecast for two or three days (which is being introduced in the management system of the electrical generation park), it is even possible to optimize the energy production to make it available during the hours when energy costs more. An accumulation system also allows production on requirement, contributing to the creation of the required margin for the power stock of the net [45].

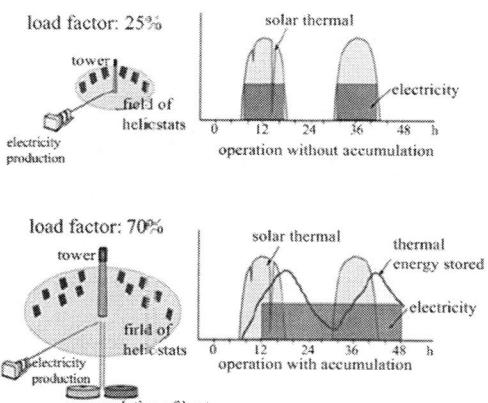

Figure 93: Production scheme with and without the accumulation system.

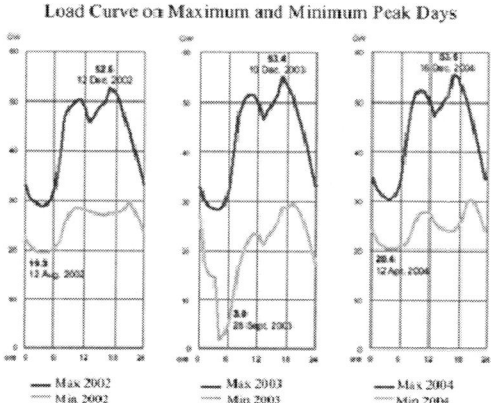

Figure 94: Loading diagrams of the Italian net in 2003.

4.4.3 Parabolic dish collector systems

This system uses reflecting panels which have a parabolic shape and track the Sun by rotating around two orthogonal axes. These panels also concentrate solar radiation towards a receiver which is installed at the focal point. High temperature heat (>650°C) is normally transferred to a fluid (helium or sodium vapour) and is then used in a motor, which is located above the receiver (see Fig. 95), where mechanical or electrical energy is directly produced. For economic reasons, concentrator dimensions do not exceed a diameter of 15 m, limiting its power to about 25–30 kW_e. With a row of these collectors, it is possible to realize systems of any size and power. An interesting application of parabolic dish collectors is the one which regards electrical energy production for small communities which are decentralized and distant. These systems have a conversion efficiency which is more than 30% (the highest efficiency among the currently existent solar technologies) [45, 50, 51]. This technology has now reached the industrial phase, mostly due to the research which has been developed in Europe, in the USA and in Australia.

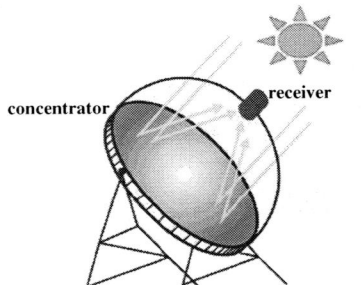 

Figure 95: Single parabolic dish collector.

Among the described technologies, this system is the one which has the highest electrical energy production cost (in 2004 costs were about 1 €/kW h); nevertheless, it is interesting for the prospects it offers concerning the drop in this cost [50]. The cost for the construction of a solar thermal electrical system which uses parabolic dish collectors in 2004 was about 7100–3700 €/kW_e with a forecast for the medium term of about 2000–1200 €/kW_e.

In the parabolic dish collectors, the thermal vector fluid can reach temperatures which can be even higher than 1000°C, and at such high temperatures it is also possible to produce hydrogen by the dissociation of water. In prospect, this is the most important reason for the interest shown in this technology: in Europe, since 2002 the *hydrogen economy* has become one of the mainstays of the EU sustainable energy policy, acknowledging the uniqueness of hydrogen both as a clean fuel and as a high efficiency energy vector [45, 53, 56].

Figure 9e: Parabolic dish collectors.

4.4.4 The use of CSP technology for electricity production

Currently, the most concrete application that CSP plants find use in is the production of electricity.

Also, in the medium to brief term, it is predicted that this will be their main application. Similar to all the other forms of renewable energy that have been introduced recently, to affirm themselves, the CSP plants must face the hard competition as regards the energy generation costs.

The presence of a growing market for 'green energy', which calculates a form of economic incentive, allows to overcome, in favourable cases and taking into consideration the growing cost of fossil fuel energy, the competition gap compared to traditional products.

To acquire market quotations it is necessary to reduce the production costs and to improve the market value of the energy produced. As for the cost reduction, two main modalities can be adopted: the reduction in the specific costs of investment and the increment of production efficiency. The improvement in the market value can be achieved by making the electrical energy production less dependent on the solar source variability. The introduction of a storage system or the use of an integrated solar-combustible fossil system is indispensable.

It must be stressed that reduction in the investment costs, improvement in the efficiency and independence from the solar source variability are contrasting aspects. Achieving a winning compromise in the course of time is very necessary, but it will necessarily leave space for specific innovations (as it happens in the automobile technology or, in a more persistent way, in the Aeolian technology) [45].

4.4.5 The future: the direct production of solar hydrogen

One of the main problems in a future sustainable energy scenario will be the large-scale production of energy at competitive costs and without the production of greenhouse gases, which will be served by transmission vectors such as electrical energy and hydrogen. Once the potential of 'energy transfer' through electrical interconnection is used up, the new hydrogen vector will allow transferring, over long distances, the potential of the primary sources from the production areas to the consumption areas, as it currently happens for the fossil fuels. Obtaining large quantities of energy to vector as hydrogen without the emission of greenhouse gases means using water as a raw material and as a primary energy source, which does not produce greenhouse gas. The hydrogen production from solar concentration systems, by means of thermochemical processes at high temperatures, promises achieving high earnings in terms of conversion, which are necessary for the process effectiveness. In fact, if presently the hydrogen production by electrolysis is the more mature process to obtain hydrogen from solar source, this process is characterized by a global yield (from hydrogen energetic content radiant energy, passing through the collection and radiation concentration, the conversion into electricity and electrolysis) of the order of 27%. Using photovoltaic conversion for electricity production, followed by electrolysis of water, we do not obtain higher yields, but we typically reach a global yield of the order of 12%. Except the costs, which are currently hard to evaluate, from the energetic point of view they are more useful than those methods in which the heat solar conversion in hydrogen happens in a direct way, based on the scheme represented in Fig. 97; in this way, theoretically it is possible to obtain global conversion yields of the order of 46%.

The thermochemical cycles, comprising oxidation–reduction reaction series that involve different natural intermediate substances, allow the cleavage of the water into hydrogen and oxygen starting from relatively elevated temperatures of heat (800–1500°C), but in solar concentration systems these temperatures are, however, achieved using high concentration systems such as towers or parabolic disk systems. This typology of the process is known since the 1970s, but only during the last few years it has become the object of renewed interest, driven more and more by the impelling environmental problems. The possibility to thermally feed such cycles by solar energy makes these production systems completely renewable and so perfectly compatible with a sustainable development strategy [45].

4.5 The ENEA technological proposal for solar electricity: the use of molten salts in parabolic collector systems

Since 2000, ENEA has undertaken research, development and demonstrative production activity in the field of solar concentration technology that aims at electricity production in a short- and medium-term perspective.

The technology developed by ENEA combines some characteristics of linear parabolic collector systems and tower systems with the aim of creating series of technological innovations that will allow going beyond the critical points of both these systems.

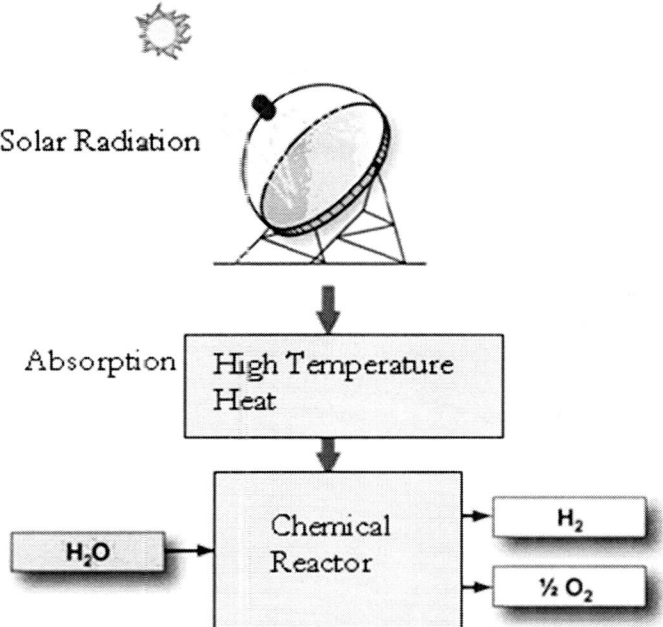

Figure 97: Hydrogen production scheme from the solar source by a thermochemical process.

In particular, the features of the ENEA technology are [45, 50, 53]:

- the use of linear parabolic collectors (because it is a more mature technology), but they are renewed compared to the traditional ones (see par. 4.5.2);
- the development of a receiver pipe capable of operating at high temperature (see par. 4.5.2);
- the use of a mixture of molten salts (made of 60% sodium nitrate and 40% potassium nitrate) that is already used in tower plants as the heat transfer fluid in place of the synthetic oil which is used in traditional linear parabolic collectors (e.g. in the SEGS);
- the presence of a thermal storage system, which was also already used in the tower systems but is absent in the traditional linear parabolic collector plants, allows storing the collected thermal energy and making it available continuously at night and during cloudy days or in case of damage to the receiving system.

The working scheme of an ENEA linear parabolic collector plant using molten salts is shown in Fig. 93.

There are two reservoirs (one 'hot' and another 'cold') that contain the mixture of molten salts, respectively, at temperatures of 550°C and 290°C. From the

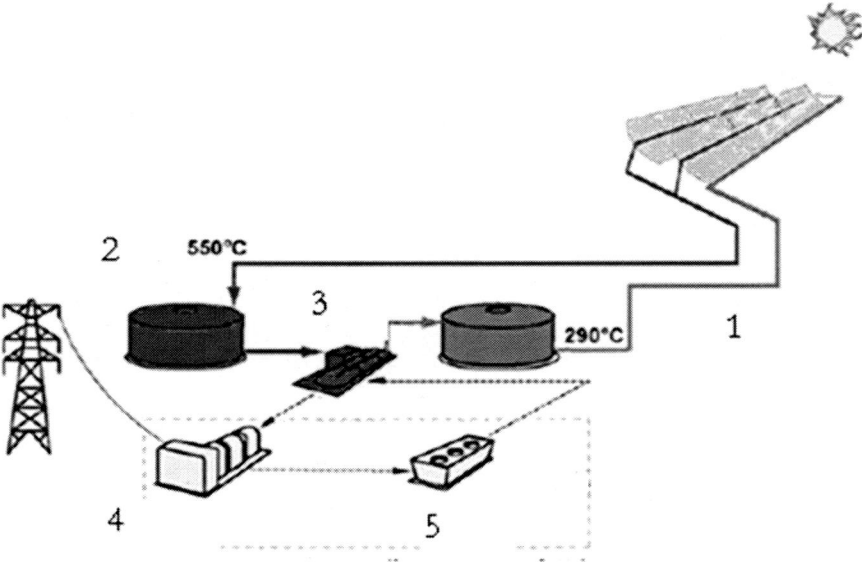

1- molten salts
2- storage tanks
3- heat generator
4- turbine and alternator
5- condenser

Figure 98: ENEA technology scheme for molten salts plant. 1: molten salts; 2: storage tanks; 3: heat generator; 4: turbine and alternator; 5: condenser

reservoirs there are two independent circuits in which the salt is circulated by appropriate circulation pumps. In the circuit of the solar field, in the presence of enough irradiation, the salt taken from the cold tank heats up to 550°C circulating inside the solar collectors and then fills the hot tank. In the circuit of the vapour generator (GV), the salt is taken from the hot tank and after having produced overheated vapour in the GV it goes back to the cold tank. The vapour produced in the generator feeds a conventional electrical energy production system.

In the limits of the storage capability, the two cycles (one relating to the solar energy consumption and the other relating to vapour production to feed the electrical generation system) are completely free, permitting electricity production which is verifiable apart from the availability of solar irradiation [45, 50, 53].

4.5.1 The advantages of molten salts

The use of molten salts as the heat transfer fluid, instead of synthetic oil, provides two advantages:

- Thermal storage can be achieved at a low cost, because the salts are economical, not toxic and have limited environmental impact even if there is an accidental outburst.

- The temperature at the exit of the solar field can be raised up to 550°C (Fig. 98), resulting in an improvement in the performance of the thermodynamic cycles involved in electricity productions. In the case of synthetic oil, the highest temperature is, on the contrary, limited to about 390°C (see par. 4.4.1).

Compared to the storage, as already seen, the cost of the fluid, especially the high danger of burning and the major environmental impact in the case of an accidental outburst, makes the realization of thermal storages using synthetic oil impracticable. On the contrary, a thermal storage system with molten salts is more useful from the cost and safety points of view, as is also proposed for oil systems (such as the two AndaSol systems with oil parabolic collectors and a capacity of 50 MW_e, in the realization phase in Spain) with the use of appropriate heat exchangers.

Figure 99: Storage reservoirs.

In this case, all the features of the molten salts are not exploitable because its temperature is limited compared with the highest temperature of the oil circuit. But by exploiting the highest temperature of the salt, we can obtain a thermal storage density of the order of 0.2 MW h/m^3, which is more than double with reference to a molten salts reservoir inserted into an oil circuit. This, combined with the low cost and the high density of the molten salts, allows achieving a cost of 15 €/kW h_t for the storage. In terms of electricity production, considering the thermodynamic conversion yield, we obtain a specific cost (cost of investment which is necessary to assure a storage thermal capacity capable of generating 1 kW h electricity) equal to 36 €/kW h_e.

This storage technology is useful compared to the other forms of storage that are used in the field of electricity production, such as the storage in hydraulic basins (pumping/turbine plants) and electrical storage with batteries, having a lower investment cost: in fact, these systems have costs (for technologies already in commercial use) of the order of 100 €/kW he and 100–1000 €/kW he, depending on the cases [45, 50, 53].

The use of molten salts produces vapour at high temperatures, of the order of 530°C, able to feed vapour cycles with high thermodynamic conversion efficiencies (42:44% against 37.6% for a vapour feed cycle, which is typical of an oil plant), without the use of a fossil fuel re-heater.

Apart from the advantages described above, the use of molten salts causes more relevant technological problems than the use of synthetic oil. The main problem is that these mixtures are liquids only at temperatures higher than 238°C and hence it is necessary to adopt technical solutions for their usage. In particular, it is necessary to have a salt fusion system and preheating electrical pipes system during the 'the first start' of the plant (when the pipes have to be filled in with salt) as well as to ensure continuous circulation of the salts in the pipes (even at night) to prevent solidification of the mixture. An alternative to the continuous circulation can be the daily filling and emptying of the circuit, but this operation is only practicable in piping limited extension plants. Another aspect that characterizes the molten salts is the necessity of adopting appropriate materials and constructive technologies for pipes and component production (particularly pumps and valves), particularly for the corrosion behaviour [45].

4.5.2 The solar collector used by ENEA

The solar collector represents the main aspect of the economic analysis that will decide the realization of a solar central plant and hence its cost and efficiency are of particular importance for the diffusion of the concentration of solar technology. For this, ENEA has planned and realized, together with the industry, an original prototype of the linear parabolic collector with the double aim of improving the techno-economic parameters and putting the national industry in a situation to produce this in series, both for the Archimedes Project (see par. 4.5.3) and for making it available in the international market in a competitive manner.

The ENEA collector, shown in Figs 100–102 comprises:

- a structure that supports the mirrors, realizing the parabolic geometry and it allows orienting them to follow the motion of the Sun;
- a series of mirrors of appropriate geometrical design;
- a motion system that is capable of making the structure rotate with the accuracy of required pointing.
- a series of receiver pipes on which the solar rays are concentrated and in which the thermal energy is given to the vector fluid.

The collector was developed in its entirety and tested using a circuit test at the ENEA centre in Casaccia. The length of the prototype collector is equal to 50 m, but the combined length of the series is 100 m. The structure must contemporaneously assure rigidity, geometrical precision and low cost. The main aspect as regards the

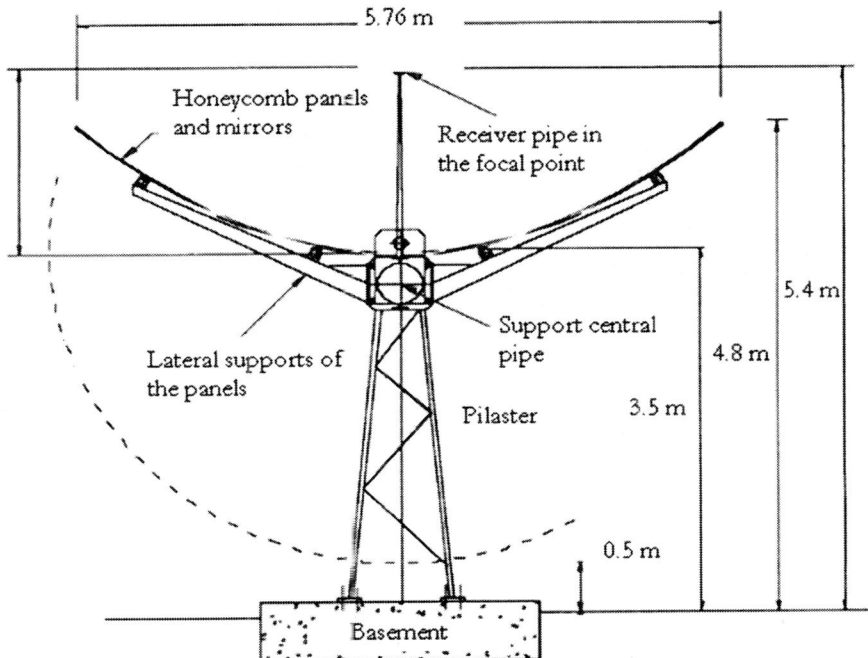

Figure 100: Solar collector scheme used by ENEA.

dimensioning of the structure is the determination of the aerodynamic loads due to the wind action. The solution adopted by ENEA, apart from presenting structural resistance, is characterized by constructive economy and simplicity of assembly. Such a solution based on a central bearing pipe and lateral supports of a variable shape (Fig. 100), heavier than similar concurrent realizations, presents its trump card in terms of the constructive reasons and in the choice of the material, which make it less expensive, easily portable, with quick installation and simple registration, within the required distance from the optical concentrator system. Despite the considerable dimensions (total length of 100 m, width 6 m and height 3.5 m from the rotation axis), after assembly tolerances of 1 mm final can be easily realized [45, 53].

The mirrors are realized with several technologies with the aim of exploring a series of alternatives to achieve a lower final cost and better mechanical characteristics compared with the traditional linear parabolic collector, which uses a thick and hot bent glass mirror. Among the alternatives examined, the better solution is the one which is based on the use of a glass mirror which is sufficiently thin (850 μm) to be cold bent until it assumes the required parabolic shape and to be applied to a conveniently shaped support panel with a structural function (Fig. 103). The support panel is made of an aluminium nucleus with a honeycomb structure, which is often 2.5 cm, and wrapped between two surface layers (leather), 1 mm thick, in composite material [45, 50, 53].

Figure 101: The molten-salt-cooled parabolic trough collector at ENEA, Casaccia, Rome.

Figure 102: Close-up of the ENEA collectors.

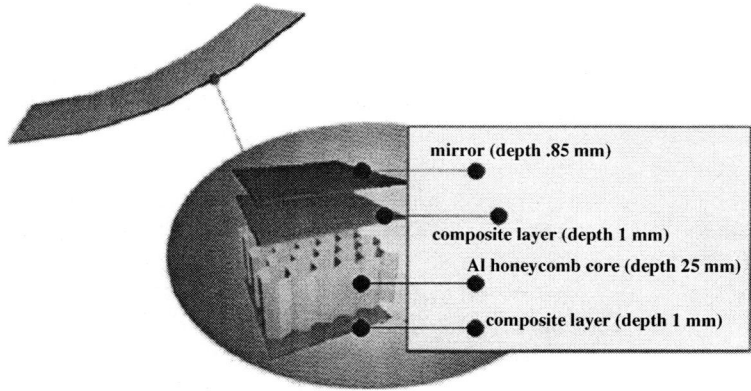

Figure 103: Honeycomb technology scheme.

The handling system is made of an autonomous oleo dynamic unit that is capable of moving the whole 100 m collector on the basis of instructions sent to the central supervision system, ensuring that the movement the Sun is followed with a precision of 0.8 mrad.

Figure 104: Collector motion system.

Figure 105: Motion system detail.

The system is able to carry the collector safely (when faced with atmospheric events such as strong wind or hail) in the presence of winds with speeds up to 14 m/s; once placed safely, the collector can resist winds with speeds of up to 28 m/s [45].

The receiver pipes (4 m long) are welded to make a line that, in the position of reference during use, must be in axis with the focal line of the parabolic mirrors. The receiver pipeline is held in position by sustaining arms equipped at the extremities with cylindrical hinges which allow the thermal expansion of the pipes when the plant is in use.

The function of the receiver pipes is to transform heat at high temperature and to transfer to the heat transformer fluid the largest quantity, reducing at least the losses of energy by irradiation towards the external environment.

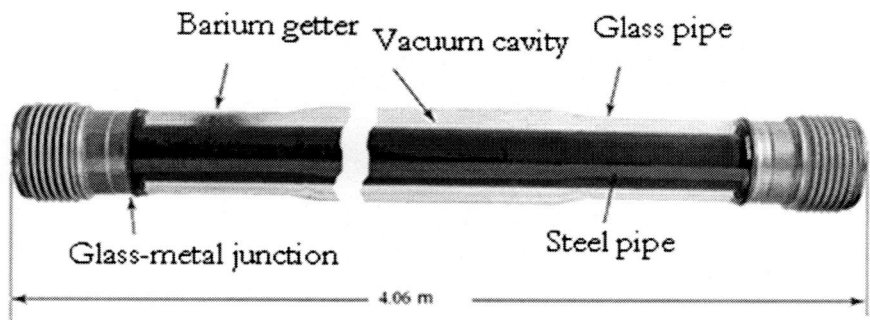

Figure 106: Receiver pipe structure.

Each receiver pipe (Fig. 106) is made of a stainless steel absorber on whose external surface is deposited, by sputtering technology, a selective spectral covering (coating) made of composite cermet material (CERMET), which is characterized by an elevated absorbance of the solar radiation and a low emissivity of heat in the infrared region. The stainless steel absorber is capped, vacuum at about 10^{-2} Pa, in an external borosilicate glass pipe that is coaxial with the receiver pipe; this external glass pipe protects the receiver pipe from the contact with air, reducing at least the thermal exchange for convention between the pipes.

On the surface of the glass pipes, an antiglare treatment is made to improve the transmittance of the solar radiation, reducing the reflected energy. The links between the glass and the steel pipes are realized with two stainless bellows (placed at the extremities of the glass pipe), which are able to compensate the differential between the two materials' thermal dilatations. To create the vacuum it is necessary to insert in the cavity between the two pipes an appropriate quantity of getter material which is capable of absorbing the gas mixture that could form in the receiver pipe.

A second material absorber, which is very reactive with air (Barium getter), is deposited on the internal surface of the glass pipe, resulting in metal colour scrubs of some cm^2. When the vacuum is created in the pipe the soaking getter saturates, the scrub becomes white, indicating the loss of heat transmission efficiency to the heat transfer fluid. The receiver pipe is the most delicate element of the solar technology, because it has to grant in time a high energy absorbing coefficient which is concentrated from the parabolic mirrors, limiting at maximum the losses by irradiation towards the environment. To achieve high reliability, there are two important characteristics:

- the capacity of CERMET to maintain almost unweathered the photo-thermal characteristics at the maximum working temperature of the coating (580°C);
- the capacity of the metal–glass junctions to resist the strains of thermo-mechanical fatigue which originate from the variability in the solar irradiation (the maximum temperature of reference in the proofs of the mechanical characterizations is 400°C).

These characteristics, peculiar to the ENEA project, have led to the development of new technological solutions because the receiver pipes which are available in the market are able to operate up to a maximum coating temperature of 400°C. In the ENEA laboratories in Portici, different CERMET made of metal and ceramic material are being planned, realized and undergoing spectrally selective characterization until the chemical composition and the optimal physical characteristics to obtain the photo-thermal characteristics required by the ENEA project are attained [45, 53]. The reference parameters of the coating developed in Portici, determined by photo-thermal characterization at the same laboratories, are:

- high photo-thermal efficiency, which means high solar absorbance (>94%) and low emissivity (<14%) up to the temperature of 580°C;
- high chemical and structural stability up to the temperature of 580°C.

4.5.3 The Archimedes Project

A fundamental aspect for the development of the ENEA technology is the realization of demonstrative applications on an industrial scale. The realization of a complete solar prototype plant for electrical energy production, linked to the national distribution net, needs the participation of public and private initiatives, as well as adequate investments. In fact, the prototype plant involves high costs because of the essential phase of learning in the setting up and use of new technologies; to be economically viable, plants of this kind need to produce more than 40 MW_e of power. But, solar plants can also be integrated with conventional thermo-electrical systems, even those with combined cycles, to improve the total amount electrical energy produced. This possibility allows using, with small changes, already working installations. So we can rely upon the electricity production system on the site and the existing infrastructure, limiting the cost for the conventional part of the plant as much as possible and focusing the investment on the innovative components of the new technology. In this case, the improvement in the power can also be widely modulated during the day, making the additional production of the solar plant to happen during the hours when the external users' demand is higher [53]. In line with what we have just stated, on 26 March

2007, Santo Fontecedro, Director of the General Division and ENEL Energy Management, and Luigi Paganetto, President of ENEA, signed, in the presence of the Environment Minister Alfonso Pecorario Scanio and the Nobel prize winner Carlo Rubbia (former president of ENEA), an agreement protocol to make the Archimedes Project operative. This project located in Sicily at Priolo Gargallo (Siracusa) represents the integration of an ENEL combined cycle thermo-electric solar plant, comprising two sections of 380 MW$_e$ each (250 MW$_e$ for the turbo gas group and 130 MW$_e$ for the vapour group) to produce a total power of 760 MW$_e$, with a thermodynamic solar plant based on the newly elaborated ENEA technology. The Archimedes Project is the main demonstrative realization of the ENEA technology and it will be the first application at a worldwide level of the integration between a combined cycle plant and a thermodynamic solar plant [58].

The choice of Priolo Gallo was made based on the following technical reasons:

- Currently, there is a considerable availability of land that is not being used of almost 60 ha in the central area.
- The site enjoys elevated insulation values, with a medium year direct solar irradiation equal to 1,748 kW h/m^2 a year.
- The vapour produced from the solar plant, having practically the same temperature and pressure characteristics as that coming from the heat recovering generator of the discharge fumes of the turbo gas, will be directly emitted into the vapour turbine of the existing central part (see Fig. 107), allowing to save

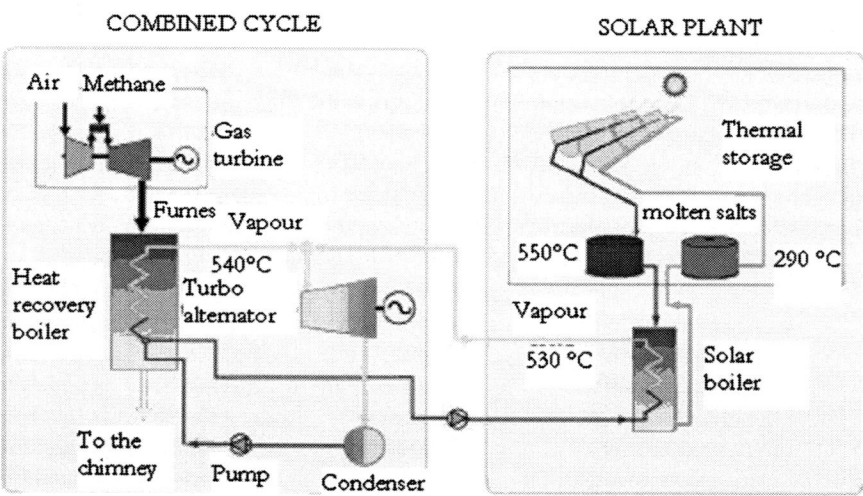

Figure 107: Integration of a solar plant with a combined cycle plant according to the Archimedes Project scheme.

Figure 108: ENEL Priolo Gargallo thermoelectric central scenery.

on the entire conventional part (avoiding the installation of a turbo alternator group and electrical instruments for the internet connection).
- The integration with ENEL central will permit the exploitation of a series of technical infrastructures and it will ease the experimental management.

The big solar plant will improve the power from the central of 28 MW_e, against an occupied area equal to 37.6 ha. The calculated electrical production net is equal to 54.2 GW_e/year, with a primary energy saving equal to 11,835 Tep and the missed emission of 36,306 t of CO_2. The global yearly medium earning (from solar energy to electricity) is equal to 17.3% [59].

The realization of a first plant module made of 60 collectors of 100 m, equivalent to about 5 MW_e, was already planned. Such a module will permit the production of [58]:

- additional electrical energy from the solar source, which can satisfy the yearly requirements of 4,500 families;
- a saving equivalent to 2,400 t of petrol;
- lower emissions of carbon oxide of about 7,300 t in a year.

Once the demonstrative plant is completed, a more concrete commercial perspective will open up, and some Italian companies have already been authorized to produce components for the emerging Spanish market. In an elevated irradiation site such as in the North African area, the ENEA technology permits, in perspective, the production of 275 GW h/year of electricity at a levelled cost equal to 4.5 c€/kW h for each square kilometre of the territory, with a saving of primary energy which is equal to 60 ktep/year and an avoided emission equal to 185 kt/year of CO_2 [59].

Figure 109: Photographic simulation of the Archimedes Project solar plant.

4.6 Conclusions

The concentration of solar technology can play a fundamental role in the future of the world's energy production, allowing the production of large quantities of electricity and hydrogen, which are completely renewable and without emission of greenhouse gases, at competitive costs. The available theoretical potential in the 'sun-belt' countries is in fact large enough to ensure a meaningful contribution to the predictable world requirements. The technological maturity regarding electricity production will be realized in the medium to brief term, and regarding hydrogen production in the medium to long term.

Especially the countries facing the southern side of the Mediterranean and the Near East countries have notable powers, with direct insulation characteristics of 50–60%, which is higher than what was found in the most favourable areas, from the southern Europe point of view. This strong insulation and the presence of large areas that are appropriate for the installation of solar concentration plants will result in energy production costs which are really lower than that in Europe.

This fact has lead to a renewed interest in proposing countries with a strong technological background, such as Germany, as candidates for ambitious development plans in collaboration with the middle-southern and Mediterranean area countries.

The presence of areas which are favourable for the concentration of solar technologies, in the southern European countries also (not only in Spain but also in Italy and Germany), has allowed the building of prototype plants to create a solid industrial base so that they can take advantage of, in addition to the energy production (especially from the manufacturing experience), the huge potential exploitation, with returns in terms of supplies for the national industries. In fact, it is evident that, being primarily a free source, the total invoice which is linked to energy production from the solar source is good for those who realize and takes care of the maintenance of production plants, and those who have the know-how are destined to exploit most of the connected businesses.

The solar concentration technology can soon be integrated, even in Italy, with other renewable technologies (Aeolian and solar photovoltaic), which will contribute to the growing European demand of 'green electricity' [45].

4.7 Solar technologies for electricity generation without light concentration

Let us now consider two other technologies which exploit the solar radiation and are applied as CSP technologies for the generation of electrical energy, but they do not involve the concentration of solar beams. This raises the possibility, in the two case that we are going to discuss, of exploiting not only the direct radiation but also the indirect radiation which, in some seasons and in some countries, has a higher energy than direct radiation.

Solar chimneys, similar to solar ponds, are not characterized by other typical temperatures, whereas the CSP technology is [7].

4.7.1 Solar chimneys/towers

Solar chimney plants allow producing electrical energy in a renewable way. They are made of a tower that is hollow inside and at the base it has a wide greenhouse, generally circular in shape that covers a notable ground surface. The greenhouse air, heated by the Sun, rises along the chimneys due to two physical phenomena (that function as the tower's 'motors'), namely:

- the air rises by floating (based on the phenomenon that hot air tends to rise high);
- the air rises due to the pressure difference between the base and the top of the tower (at the top of the chimney the pressure is lower and so the air is 'backwashed' towards the top).

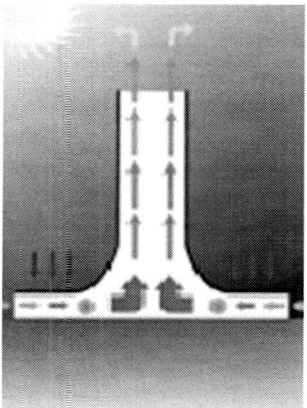

Figure 110: Solar chimney.

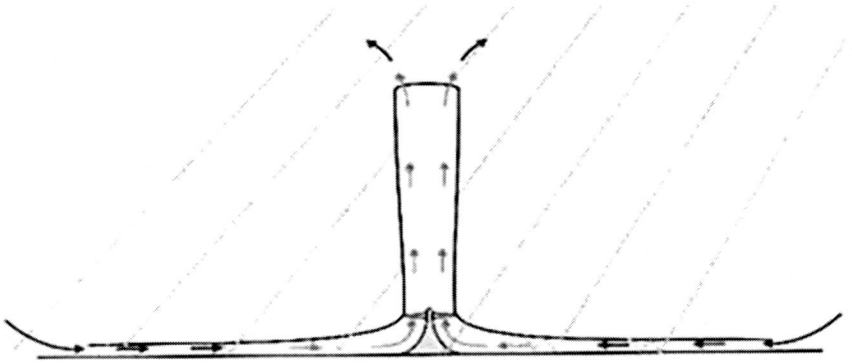

Figure 111: Principle of the solar chimney.

As it rises in the chimney, the hot air accelerates until it reaches a speed of 70 km/h. This flow of air rotates a series of turbines placed at the internal base of the chimney to generate electricity: the turbines transform the kinetic energy and the air potential into electrical energy, as every Aeolian blade. The procedure is made easier from the absolute constancy both in direction and in intensity of the speed vector.

The heat collector in this case is the greenhouse. It can have plastic or glass covers. From the pilot plant at Mazanares (Spain, Fig. 112) we can see that the glass is better because it is more resistant to bad weather. We also observed that if the height of the cover progressively improves towards the centre, the radial flow of the speed is enhanced. The performance directly depends on the chimney height. For this reason, in the current plans, they plan to build chimneys of 1,000 m height.

Figure 112: Mazanares solar tower.

The main feature that makes the solar chimney/tower particularly interesting is its capacity to work without wind also, 24 hours, 7 days, generating a peak of energy during the hotter days of the year when there is a consumption peak.

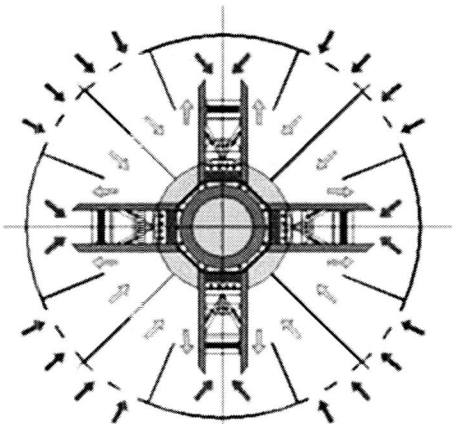

Figure 113: Solar chimney.

The plant can also work at night, due to the 'pressure gradient' (i.e. the pressure differential) and, secondly, due to the ground covered by the greenhouse, which heats itself during the day and releases the stored heat during the night. We can easily improve the thermal capacity of the floor by putting a water layer in the greenhouse or using an appropriate arrangement containing water elements that store the heat and release it at night. Obviously, water must be contained and kept; it must not evaporate; otherwise, it consumes the thermal energy absorbed [7, 61, 63].

Among the most ambitious project in terms of dimensions is, without doubt, the solar chimney/tower that to be built in the county of Wentworth in New South Wales, Australia. Figure 114, where the greenhouse cover elements are considered the solar panels, shows the scheme for this project. The numbers of the initial project are as follows [60]:

- The greenhouse should cover an area of about 25,000 acres, which is equal to 5 km^2.
- The central tower will be 3,280 feet high, corresponding to 1 km, which would make it the tallest building in the world.
- Inside the tower 32 turbines each of 6.25 MW are placed; every rising hot air motion is estimated to have a maximum speed of the order of 35 miles/hour (<60 km/h); the solar tower will have a total capacity of 200 MW, which is enough to feed almost 200,000 houses.
- The generation of 200 MW of power would allow saving, depending on estimates, between 750,000 and 900,000 t of CO_2 per year.

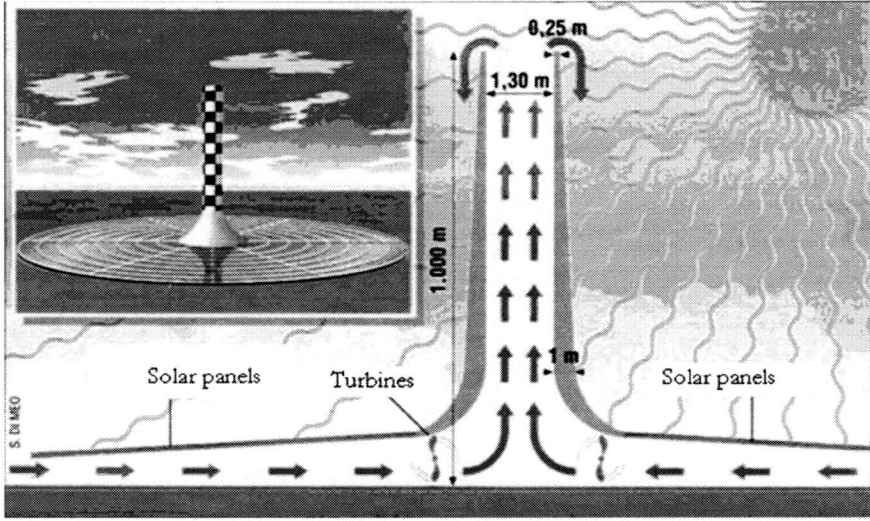

Figure 114: Australian solar chimney scheme.

Currently, the project is in the final stage in terms of its feasibility, particularly regarding the economic aspects. In this step, the Guinness dimensions of the initial project have been reduced:

- The use of innovative materials has allowed reducing the height of the tower to 650 m without losing power.
- The power has been reduced to 50 MW.
- At the moment, it is not possible to know the final dimensions of the tower, but it is reasonable to assume that at such levels it should have a height of at least 450 m.

From the technical point of view, the project was already validated, because for 7 years (from 1981 to 1988) a pilot project of 50 kW power was operative at Mazanares. Conceptually, it is not a new technology, but at the moment of its birth, when an oil barrel cost 15 dollars, it did not provoke any particular interest, contrary to the situation today. In fact, the present high price of crude oil and the necessity of reducing greenhouse gas emissions are pushing many countries towards more convenient and cleaner energy sources such as the solar chimney/tower [60].

The highest and most sophisticated solar chimney/tower (750 m) in Europe will be realized at Fuente del Fresno, in the Spanish region of Mancha. This colossal solar system will have a power of 30 MW. This plant will provide electrical energy that is equal to the requirements of 120,000 people and at the same time we will avoid putting into the atmosphere 78 t of CO_2 that will be generated from 140,000 oil barrels that could produce the same energy in a year. The construction of this structure will start in 2007 and it will be finished in three years; it will cost 240,000,000 € and it will occupy 350 ha covered with a 3 km diameter crystal panel. Exploiting the greenhouse effect principle, the overheated air will rise along the tower height, actuating 24 turbines that will produce electricity. A system of storage pipes filled with a gel keeps heat and allows the generators to produce energy even at night and during periods of scarce insulation. The tower has an estimated shelf life of 60 years [65].

4.7.2 Solar ponds

The term 'solar ponds' is used to describe a mass contained in a basin of water that also absorbs the solar incident energy and stores it in its interiors. To obtain this performance, three other basic kinds of solar lakes can be named, and they are identified by the terms: salinity gradient solar lake, gel pond and, finally, shallow solar pond. Among the three, the first is the one whose technique was realized for the totalities of the realizations and the management of the physical working studies. This kind of solar lake is realized putting in the bondage a solution of salt in water, e.g. sodium chloride, using filling techniques that allow establishing a growing salt concentration with the depth until the saturation at the bottom layer. Effectively, in the vertical section of the basin (see Fig. 115), which is generally deep 2–3 m, we can find three characteristically superimposed layers: the first layer is high and very slender, it is composed of water with a little quantity of salt (0–35 g/l); the central layer, where we can observe a linear salinity variation; and, finally, the homogeneous and salt saturated bottom layer (200–250 g/l).

Let us now analyse the difference between a normal water basin and a solar pond. In the first case, the solar energy heats water (exposed to the Sun), which,

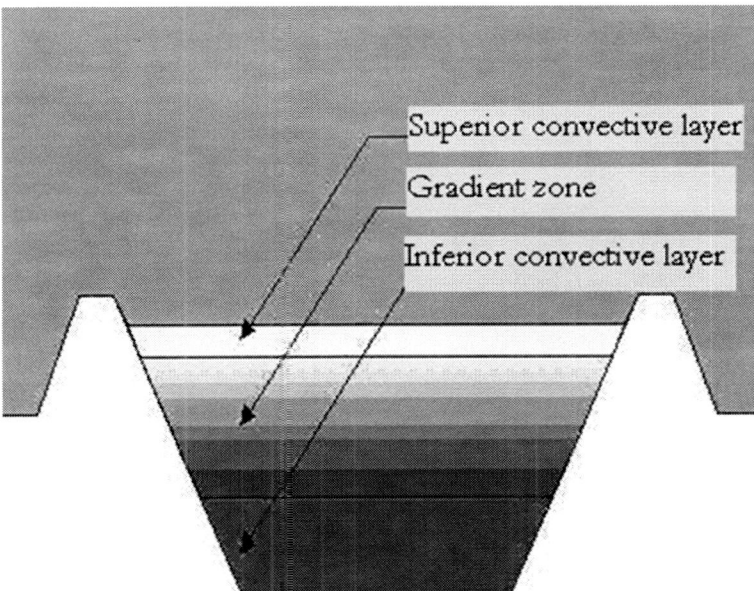

Figure 115: Solar pond scheme.

however, tends to lose this heat. Indeed the water heated by the Sun expands and tends to move higher and higher as it becomes less dense. Convective motions are established and the superficial water is always hotter than the deep water; it rapidly evaporates cooling and giving heat to air. The cold water, which is heavier, moves towards the bottom. In this way, a water basin keeps a relatively low temperature in the deep bottoms and, as it is more radiated, it raises the circulation speed of the water and intensifies the evaporation. But if a system in which the mass of water has a layer shaped salinity is created, with the highest value at the bottom and the lowest value at the surface (solar pond), the convective motions are inhibited. In fact, the hot water specific gravity and high salinity are anyhow bigger than that of the modest salinity cold water, so heat is trapped at the bottom of the solar pond. The absence of convective motions inhibits the mixing of high salinity hot water with the superficial one. The superficial layers of salinity only increase diffusion and this happens over very long periods (years) and so bigger the solar pond spare part time that has to be fed to equalize the losses of evaporation.

When the solar radiation incident on the solar pond surface penetrates through the transparent solution mass, it is absorbed at the bottom and the produced heat transmits itself to the solution for convention. Following mass ascent and energy transfer that could lead to the dissipation of the heat at the surface, it finds a barrier in the interface with the salinity gradient layer and the heat is stored in the pickle at the bottom (where the temperature can also reach 100°C). In fact, the water in the salinity gradient area cannot rise because the water in this layer has a lower salinity content and so it is lighter; for the same reason, the water in the higher layers cannot go down because the water in the lower layer has a salinity content which is lower and heavier and even if its density wanes with the increase in temperature, it is always denser than the higher water layers. The intermediate layer acts as a transparent thermal isolator that allows the stored heat in the lower convective layer to be extracted with thermal exchange techniques and to be used for thermal purposes [66–68].

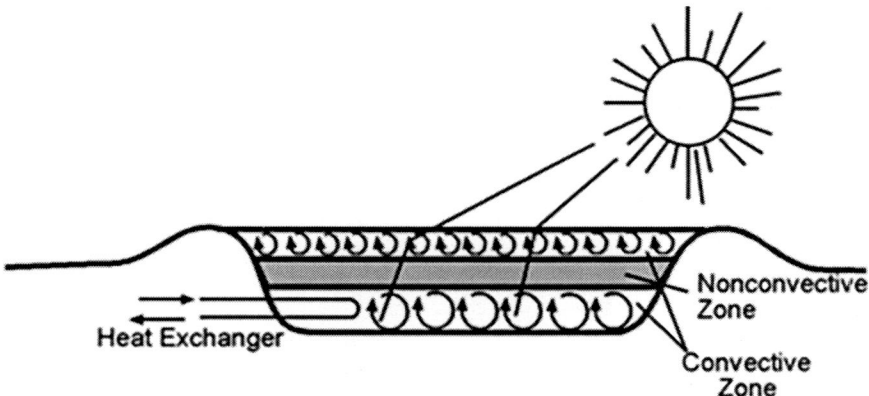

Figure 116: Working scheme of a solar pond.

Solar ponds are mainly used as energy sources which are appropriate to feed the processes of [7, 66]:

- electrical energy production using organic fluid Rankine cycles; the electrical production yield of the system is very low, but the cost of the storage plant is contained;
- brackish water desalinization;
- agricultural greenhouses and habited environmental heating;
- vegetable drying.

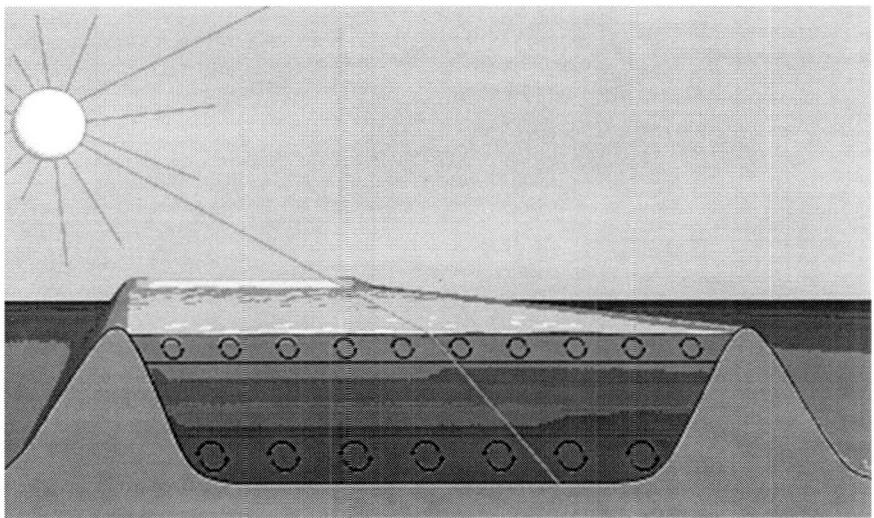

Figure 117: Convective motion scheme in a solar pond.

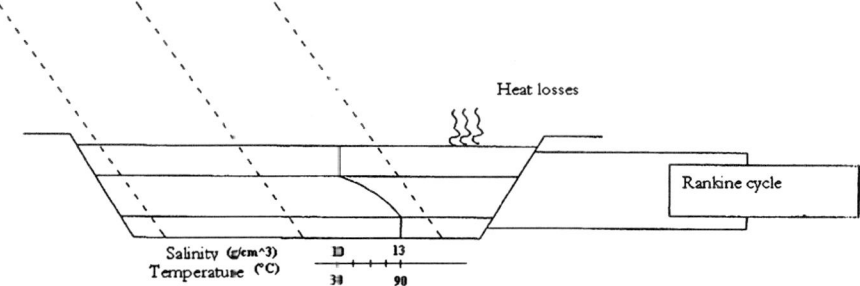

Figure 118: Principle of the solar pond applied for electrical energy production.

A solar pond can be built using normal intervention techniques used by the building industry, such as digging the basin, covering the basin with an impermeable membrane and building the structures for housing the devices used for extracting and producing heat. In this way, large heat collection surfaces can be realized, up to thousands of square metres in area with costs for unit area lower than the cost of every other methodology of solar energy exploitation. The big mass for collection and the thermal isolation capability characterize the solar pounds: they can preserve the thermal energy for long periods (seasons) without registering sensible brine temperature decreases.

The construction of a solar lake, in terms of the surface unit, can vary with the basin catchment area. The estimated unitary costs for building different size lakes are listed below:

- surface of 2,000 m^2, cost: 150 €/m^2;
- surface of 20,000 m^2, cost: 95 €/m^2;
- surface of 200,000 m^2, cost: 70 €/m^2.

Figure 119: Solar pond at El Paso (Texas).

General bibliography and consulted websites – Part I

[1] Cucumo, M.A., Marinelli, V. & Oliveti, G., *Solar Engineering: Principles and Applications*, Pitagora Editrice: Bologna, 1994.
[2] Comini, G. & Cortella, G., *General Energetic*, SGE Editrice: Padova, 1996.
[3] Schibuola, L. & Cecchinato, L., *Solar Active and Passive Systems in the Buildings*, Società Editrice: Esculapio s.r.l., 2005.
[4] Magrini, A. & Ena, D., *Active and Passive Solar Systems*, EPC LIBRI, 2006.
[5] Battisti, R., Corrado, A. & Micangeli, A., *Solar Thermal Systems: Hot Water with Solar Energy*, Franco Muzzio Editore, 2005.
[6] Paushinger, T., Mènard, M. & Schulz, M., *Thermal Solar Plants: Book for the Planning and Building*, 2003, www.ambienteitalia.it
[7] Bartolazzi, A., The Renewable Energies, Ulrico Hoepli Editore S.p.a., 2006
[8] Grillo, N., *Thermal Systems Feed with Solar Energy*, Geva Edizioni: Roma, 2003.
[9] ISES ITALIA, *Thermal Solar: Guide for Designers and for Installers*, Foundation IDIS, Città della Scienza, ISES ITALIA, 2004.
[10] Petrarca, S., Cogliani, F. & Spinelli, F., *The Global Solar Radiation at the Soil in Italy*, ENEA, 2000.
[11] European Communities Commission, *Solar Radiation European Atlas*, 1984.
[12] www2.minambiente.it
[13] www.enel.it
[14] www.rinnovabili.it
[15] www.habitat-energy.it
[16] Teaching material of the course *Technology and Economy of the Energetic Sources* AA 2004/2005, Prof. R. Basosi, Corso di Laurea in Chimica, Università Degli Studi di Siena, www.chim.unisi.it
[17] Active solar thermal, www.isaac.supsi.ch
[18] www.ecorete.it
[19] www.accomandita.it
[20] "FER solare termica", www.fo.camcom.it
[21] www.tinox.com
[22] www.oikos3l.it
[23] www.enerecosrl.com
[24] www.cullwater.com
[25] www.energoclub.it

[26] www.solarenergy.ch
[27] www.viessmann.it
[28] www.schott.com
[29] Marchini, E., *Vacuum Pipes Collectors Technology*, http://web.taed.unifi.it
[30] www.cmgsolari.it
[31] www.olivarimpianti.it
[32] www.edilportale.com
[33] Teaching material of the course of *Environmental Technique Physique*, Prof. L. Santarpia, C.d.L. Engineering for the Environment and the Territory AA.2006/2007, Università degli Studi di Roma "La Sapienza", http://pcfite.ing.uniroma1.it
[34] Croci, A. & Prosperi, M., *Dossier Solare Termico*, Progetto RES & RUE Dissemination, www.poweron.ch
[35] Tognon, E., *Il solare termico per i condomini*, atti del convegno *L'efficienza energetica nei condomini*, 2005, www.veneziaenergia.it
[36] www.lafabbricadelsole.it
[37] www.carlieuclima.it
[38] www.srb-it.com
[39] www.caleffi.it
[40] www.ediliziainrete.it
[41] www.ilportaledelsole.it
[42] IEA SHC – TASK 33, IEA SolarPACES – TASK IV, *Medium Temperature Collectors*, 2005, www.iea-ship.org
[43] IEA SHC – TASK 33, IEA SolarPACES – TASK IV, *NEWSLETTER No. 2 – Dicembre 2005*, www.iea-shc.org
[44] IEA SHC – TASK 33, IEA SolarPACES – TASK IV, *NEWSLETTER No. 1 – Dicembre 2004*, www.iea-ship.org
[45] Falchetto, M., Il Programma ENEA sull'energia solare a concentrazione ad alta temperatura, 2005, www.enea.it
[46] ENEA, Calore ad alta temperatura dall'energia solare: una tecnologia innovativa per un'energia pulita, disponibile con continuità e ad un costo competitivo, 2004, www.enea.it
[47] ENEA, Rapporto energia e ambiente 2003: le fonti rinnovabili cap. 7, www.enea.it
[48] German Aerospace Center (DLR), MED-CSP Concentrating Solar Power for the Mediterranean Region, Final Report, 2005, www.dlr.de
[49] ENEA, *Relation Energy and Environment 2005: Analyze*, cap. 6, www.enea.it
[50] Giaggioli, W., *Solar Thermal Technologies*, www.fnlevento.it
[51] www.isesitalia.it
[52] ENEA, Thermodynamic Solar Project. High Temperature Heat Production through Concentration Solar Systems, 2004, www.enea.it
[53] Mazzei, D., The Concentration Solar Technology and the Innovations ENEA, 2003, www.bologna.enea.it
[54] The Solar Thermodynamic: Existing Technologies, http://ente.provincia.parma.it

[55] www.ambientediritto.it
[56] ENEA-ENEL Heat and High Temperature from the Solar Energy, www.enel.it
[57] Press Release, ENEL ed ENEA: Starting of Archimedes, Solar Plant at High Efficiency, http://titano.sede.enea.it
[58] The Concentrating Solar Power – Global Market Initiative www.solarpaces.org
[59] http://clima.meteogiornale.it
[60] www.lifegate.it
[61] http://ilprofessorechos.blogosfere.it
[62] www.enertop.it
[63] http://associazionetcs.it
[64] *Ilsoleatrecentosessantagradi* n. 6, giugno 2006, www.ilsolea360gradi.it
[65] *Ilsoleatrecentosessantagradi* n. 1, gennaio 2002, www.ilsolea360gradi.it
[66] Leonardi, E. & Rosa-Clot. M., *Il solar pond di Cagliari: analisi tecnica*, www.crs4.it
[67] Leonardi, E. & Rosa-Clot, M., Dissalatori solari: un progetto ecologico ed economico per superare la crisi idrica in Sardegna, www.csr4.it;
[68] Bobbio, V. & Rosa-Clot, M., L'acqua dal Sole: tecnologie per la produzione d'acqua sfruttando l'energia solare, www.regioneambiente.it
[69] ENEA, Clima ambiente energia. La ricerca per un nuovo equilibrio. L'energia fotovoltaica, www.enea.it
[70] La fisica del processo fotovoltaico, http://electroporta.net
[71] Gelleti, R., *La tecnologia fotovoltaica: stato dell'arte e potenzialità di impiego nei processi produttivi* (a cura di Area Science Park), CETA Centro di Ecologia Teorica e Applicata, http://www.area.trieste.it
[72] Castello, S., Enea, *Aspetti tecnici ed economici della tecnologia fotovoltaica*. http://www.sistemifotovoltaici.com/sistemifotovoltaici.pdf, 1996.

PART II

Biomasses energy

CHAPTER 3

Biomasses identities

1 Introduction

The term biomass encompasses a large number of materials of an extremely heterogeneous nature. We can state that everything that has an organic matrix is a biomass. Plastics and fossil materials have been excluded, even though they belong to the family of carbon compounds, because they do not have anything in common with the characterization of the organic materials discussed here. In scientific terms, the word biomass includes every kind of material of biological origin; it is so linked to carbon chemistry which directly or indirectly derives from the chlorophyllian photosynthesis.

The biomass is the most sophisticated storage of solar energy. In fact, through the photosynthesis process vegetables are able to convert the radiant energy into chemical energy and to stock it as complex molecules with high energy content. For this reason, the biomass is considered renewable and unexhaustive, if appropriately used as a resource; that is, if the use tax of the same does not exceed the regeneration capacity of the vegetable forms.

The biomass is also an energy source that considers as neutral the aim of the greenhouse gas emissions increment. In fact, vegetables, through photosynthesis, contribute to the subtraction of atmospheric carbon oxide and carbon fixation in the textures (a total of 2×10^{11} tons of carbon are fixed in a year, with an energy content of the order of 70×10^3 MTep, which is equivalent to ten times the world's energy requirements).

The quantity of carbon oxide released during the decomposition of biomasses, if it happens both naturally and through energy conversion processes (even if it is through combustion), is equivalent to that absorbed during the growth of the same biomass.

There is no contribution to the increase of the CO_2 level in the atmosphere. Therefore, in this case the improvement in the quotation of the energy produced using biomasses, rather than fossil fuels, can contribute to the reduction of the CO_2 that is emitted into the atmosphere. For this reason, the use of biomass for energy applications is considered one of the priorities of the post-Kyoto development policies [2, 4–6].

Legislative Decree no. 387 of 29 December 2003 defines biomass as 'the biodegradable part of the products, wastes and coming from agriculture residuals

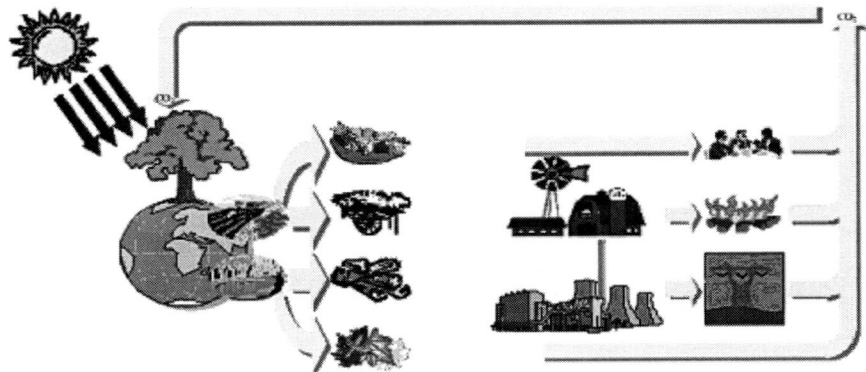

Figure 1: Organic matter cycle.

(comprehending vegetable and animal substances) from the forestry and from the connected industries, as well as the biodegradable part of the urban and industrial wastes.'

There are other terms which are associated with the term biomass, and they are now commonly used in the renewable energy sector, such as 'bio-fuel', which generally means 'every organic substance that is different from petrol, natural gas, carbon or from their derivatives, and which is usable as combustible', and 'bio-energy", which represents the energy produced from biomasses [7].

2 Definition and classification

The definition of biomass which is taken from Directive 2001/77/CE, and acknowledged at a national level from the Legislative Decree no. 387 of 29 December 2003, reunites a wide class of materials of vegetable and animal origin that also includes rubbish. More simply, the biomasses which are appropriate for energy transformation, if it happens directly using the biomass or prior to changing of the same in a solid, liquid or gaseous fuel, can be divided based on their source into the following compartments [2]:

- forest and agro-forest compartment: forest-cultural or agro-forest activities and operations residuals, use of the coppice, etc.
- agricultural compartment: farming residuals which come from the agricultural activities and the dedicated lignocelluloses species cultures, oil bearing plants for the extraction of oils and their transformation into bio-diesel, alcohol-producing plants for bio-ethanol production;
- zoo technique compartment: livestock sewage wastes for the production of biogas;
- industrial compartment: coming from wood or wood product industries and paper industries, as well as agricultural and food industry residuals;
- urban rubbish: maintenance of the public green and urban solid rubbish maintenance operations.

The term biomass groups materials that differ from each other in terms of chemical and physical characteristics. They can have multiple uses on the energy production front. Generally, as we will see hereafter, it is possible to group transformation processes into different categories: the processes of biochemical conversion, which allows the gain of energy through chemical reactions due to the presence of enzymes, fungi and other micro-organisms that form, in particular conditions where the biomass is held; the processes of thermochemical conversion has, as it basis, the heat action that allows the development of chemical reactions which are necessary to transform matter into energy. The factors that favour the choice of one of the two processes are the carbon/nitrogen (C/N) ratio and the grade of humidity at the time of collection: when the C/N ratio is lower than 30 and the humidity content exceeds 30%, biochemical processes are generally used; on the contrary, thermochemical processes are more suitable [2–5].

3 Origin and nature

3.1 The forest and agro-forest behaviour

The forest residuals, resulting from the different kinds of forest-cultural intervention, are commonly indicated as forest biomasses. The interesting operations for the sample of forest biomass aimed at energy production purposes include forest-cultural interventions in woods which are controlled both by high fores (which is applied when the wood comprises plants that are allowed to grow until maturity) and by coppice (where the growth of the plant is interrupted with periodical cuts). In the first case, a typical operation is the sample of the lower sorting, generally left in the wood after the cut of the major forest sorting (truncate with a diameter bigger than 18 cm) for commercial uses. The wood derived from the inserted cut material (interventions which are applied to young fores or to replenishing fores to improve the stability and regulate the specific composition) represents a further source of supply. Another important forest biomass source is represented by the coppice woods: the Italian coppice, in fact, is mainly destined for the production of fuel biomass and agricultural use posts.

Another source of supply for wooden biomasses is represented by the materials of agro-forest origin that are derived from forestation activity in agricultural range biomass. In this case, the usable biomass sources for energy production purposes are derived from wooden cultivation commercially used residuals, to the linear formations uses (e.g. hedges, rows and little woods) as well as wood formations uses, which are dedicated to agricultural uses (in this last case we mainly refer to the poplar culture) [2, 7, 12]. The physical characteristics of the wooden biomasses which are relevant on the energy production front are the grade of humidity and the density which, with the material's chemical composition, affect the calorific power of the wood.

The calorific power expresses the quantity of heat that is released during the complete combustion of the weight unit or in fuel volume. There are lower calorific power (LCP) and higher calorific power (HCP) fuels; depending on the hydrogen combustion that is eventually present in the fuel water, we consider the vapour or the liquid state. The difference between the two kinds of calorific power corresponds to the vaporization heat of water that is formed during combustion.

Practically, the LCP fuels are always of interest because the fumes are always discharged at a temperature where the water is present as vapour. It is expressed in MJ/kg and, for convenience, also in kcal/kg.

The humidity is a variable that assumes considerable importance because, in addition to the combustion mechanisms, it influences the chemical characteristics of the wood and its specific weight. The quantity of water that is contained in the material varies as a function of many factors such as the species, age and the plant part that is considered (trunk, branches, etc.). The humidity expresses the quantity of water (free or linked) that is present in the wood; it is expressed as a percentage in terms of both the dry weight and the fresh wood weight; in the first case, we look at the water content as an absolute value and in relation to its anhydrous mass.

$$U(\%) = [(M_i - M_a)/M_i] \times 100$$

where M_i is the exact wood mass and M_a is the mass of dry wood.

This method is the most frequently used method to determine the humidity [2, 8].

The most common wooden combustible quality indicator is represented by the density (mass for unit volume, measured in kg/m^3). It is directly proportional the wooden calorific power. The density varies as a function of factors such as the seasonal conditions, the species (the most elevated in the broad-leaved species and in the conifers), the age, the considered part and the form of the wood government. The density can be calculated by considering the wood in the fresh state or in the dry state; in the first case, it generally varies from 360 to 810 kg/m^3 [2, 8, 9]. The chemical composition of the wood is one of the main analytical characteristics for the forest and agro-forest biomass qualification on the energy production front.

The main polymers which make up the wooden biomass are [2, 8]:

- lignin, which gives rigidity to the plant (reinforces the cellular wall), is present in percentages that vary from 20% to 30% of the dry weight and has a high calorific power (6,000 kcal/kg approximately).
- cellulose, which is the main wood component (constitutes 50% of the weight) and has a calorific weight of approximately 3,900 kcal/kg;
- hemicelluloses, which are present in the cellular wall of the plants, in the free spaces left from cellulose. It constitutes from 10% to 30% of the wood and it has a more contained calorific power.

Compared to its elementary composition, wood is almost entirely made of carbon (49–51%), oxygen (41–45%) and hydrogen (5–7%). It is also composed of, even if in reduced quantities, nitrogen (0.05–0.4), sulphur (0.01–0.05) and other mineral elements that make up the cinders (0.5–1.5%) [10, 11].

The quantity and especially the relationship among the elements that make the biomass are very important to verify its value as combustible. In particular, the relationship between hydrogen and carbon and between oxygen and carbon are also really important, as well as the quantity of nitrogen and cinders; generally, a high carbon and hydrogen content determines a high calorific power, whereas elevated oxygen, nitrogen and cinders has an opposite effect [8].

Table 1: Main wooden biomass chemical–physical characteristics (ds, dry substance) [2].

Composition	
Cellulose	50% of ds
Hemicellulose	10–30% of ds
Lignin	20–30% of ds
Physical and energetic characteristics	
Humidity	25–60% on t.q.
Mass density	800–1,120 kg/m^3
LCP (considering a humidity of 12–15%)	3,600–3,800 kcal/kg

The biomass which comes from wood is sold in the market in very different coal sizes for shape and humidity grade. In some cases, it starts with the production of denser forms (pellets, briquettes, which are analysed in par. 4.1.3 and 4.1.4). The most common coal sizes are wooden stub-pipes (used in the rural or mountain environments) and chips (par. 4.1.2).

The wide availability of the source at a national level makes the exploitation of forest biomasses for energy production interesting. However, we have to face the logistic difficulties that are linked to biomass retrieval (e.g. the presence or not of an exploitable viability from the common collection and transport tools); especially in mountainous regions, in fact, the woods are not always easy to reach [2].

3.2 The agricultural compartment

In the combustible production from the biomass, the agricultural compartment has and it will have a more and more relevant role. This compartment, in fact, gives a great number of materials that are applicable for energy production (residual products that are derived both from other cultivations and from specialist cultivations that are dedicated to the production of combustible biomass materials). The main products of the agricultural sector are [2]:

- wooden cultural residuals, which come from vineyard and orchard management;
- composite nature cultural residuals which come from cereals and sowable cultivations;
- lignocelluloses from woody and herbal dedicated cultures biomasses;
- culture products of the oil culture (sows, etc.) for the production of vegetable oils and bio-diesel;
- the products of the alcohol-producing cultures (tubercles, prills, etc.) for the production of bio-ethanol

3.2.1 Agricultural residuals

The agricultural residuals include the set of by-products which are derived from the cultures cultivation, and they are generally for an alimentary purpose; otherwise, they are not usable or have alternative and marginal uses. The residuals that come from this compartment show physical and energy characteristics that, together with economic barriers (collection costs, low density for unit surface), do not make them easily applicable for energy production. For this purpose, the following can be applied:

- straws of autumn-winter cereals (soft wheat and hard wheat, barley, oats, rye);
- stocks, corncobs and maize sculls;
- rice straw;
- vine shoots of vine pruning;
- slash of orchards pruning;
- olive branches.

Table 2: Main chemical–physical cultural residual characteristics [2].

Agricultural subproduct	Collection humidity (%)	Medium production (t/ha)	Report C/N	Cinders (% in weight)	LCP (kcal/kg ds)
Soft wheat straw	14–20	3–6	120–130	7–10	4,100–4,200
Hard wheat	14–20	3–5	110–130	7–10	4,100–4,200
Autumn-winter other cereals straw	14–20	3–5.5	60–65	5–10	3,300–3,400
Rice straw	20–30	3–5	60–65	10–15	3,700–3,800
Maize stocks	40–60	4.5–6	40–60	5–7	4,000–4,300
Corncobs and vine shoots	30–55	1.5–2.5	70–80	2–3	4,000–4,300
Vines vine shoots	45–55	3–4	60–70	2–5	4,300–4,400
Olive branches	50–55	1–2.5	30–40	5–7	4,400–5,400
Fruit residuals	35–45	2–3	47–55	10–12	4,300–4,400

Despite the cultural residuals representing an energy source which is easily accessible, it is necessary to consider some limitations (low productivity for unit surface and chemical composition of the biomasses) which are linked to their exploitation: the quantities of agricultural residuals, which are available for a unit surface, are relatively modest and it can make the collection disadvantageous and inconvenient and also the removal and transport of the biomass to the thermal central; relative to the chemical composition of the agricultural residuals it is necessary to underline that an elevated cinder content increases the danger of formation of wastes with a damage for the burnings and also increases the particulate emissions [2].

3.2.2 Dedicated cultures

The term 'dedicated cultures', or 'energetic cultures', refers to the cultures prepared with the aim of producing biomass destined for preparing electric and/or thermal energy.

There are three main dedicated cultures:

1. cultures from lignocelluloses,
2. oil cultures and
3. alcohol producing cultures.

Table 3 lists the main usable species and the bio-fuel obtainable from them.

Table 3: Usable for energy cultivation and their characterization species [2, 7].

Species	Production cycle	Intermediate product	Transformed product
Kenaf	Herbaceous annual	Fibre	Chips
Hemp	Herbaceous annual	Fibre	Residual bundles
Miscanthus	Herbaceous annual	Fibre	
Common reed	Herbaceous long term	Fibre	
Fibre sorghum	Herbaceous annual	Fibre	
Cardoon	Herbaceous long term	Fibre	
Measly	Herbaceous long term	Fibre	
Robinia	Wooden long term	Wood	
Broom	Wooden long term	Wood	
Eucalyptus	Wooden long term	Wood	
Willow	Wooden long term	Wood	
Poplar	Wooden long term	Wood	
Rape	Herbaceous annual	Oilseed	Vegetable oil
Sunflower	Herbaceous annual	Oilseed	
Soy	Herbaceous annual	Oilseed	
Ricin	Herbaceous annual	Oilseed	
Saf flower	Herbaceous annual	Oilseed	
Sugar reed	Herbaceous annual	Rhizome	Sugars/alcohols
Sugar sorghum	Herbaceous annual	Stem	
Topinambour	Herbaceous long term	Tubercle	
Maize	Herbaceous annual	Granel	
Wheat	Herbaceous annual	Prills	

3.2.2.1 Lignocelluloses biomass cultures The lignocellulose cultures include the herbaceous or wooden species which are characterized by biomass production that is mainly composed of lignin and/or cellulose substances. The cultures are divided into three groups: annual herbaceous cultures, long-term herbaceous cultures and arboreal cultures [2, 7].

Annual herbaceous cultures They include herbaceous species which are characterized by a yearly life cycle. The most interesting ones are generally the sorghum, in addition to maize, kenaf, reed, etc. Such cultures do not occupy the ground permanently, allowing farming in rotation cycles. They can be also cultivated depending on the traditional set-aside kept in rest grounds [2, 7].

Figure 2: Kenaf, reed and fibre sorghum.

Herbaceous long-term cultures The number of herbaceous long-term exploitable species for the production of lignocellulose biomasses is really large. The most important ones are the common reed, the Miscanthus, and the measly. Such cultures, against a considerable impact on the organization of the farm holding (they occupy the ground for 10–15 years) and a high system cost, permit the production of a considerable quantity of biomass for more years and at low additional costs (compared to the annual species). They also require few fertilizers and parasiticides [2, 7].

Figure 3: Common reed, measly and miscanthus.

Arboreal cultures The energy producing wooden cultivations comprise species selected for their high yield of biomass and for their capacity of rapid growth after a cut. Usually such cultures show brief coppicing turns (2–3 years) and a high density of plants (6,000–14,000 plants/ha). In this case, we speak about short rotation forestry (SRF). Generally, in SRF, specific clones are appropriately selected to be used and the coppicing of the plants, annual or two-year, is completely mechanized using appropriate wood-chipping machines (see par. 4.1.2). The most interesting brief turn arboreal cultivable cultures are willow, poplar, rocinia, eucalyptus, broom (shrub) [2, 7].

Table 4 lists the physical and energy characteristics of the main vegetable species.

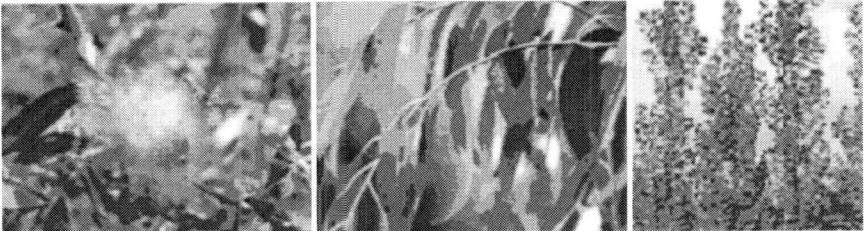

Figure 4: Willow, eucalyptus and poplar.

Table 4: Productive and energetic parameters of the biomass from dedicated cultures [2].

	Fresh substance production (t/ha year)	Collection medium humidity (%)	Dry substance production (t/ha year)	LCP (kcal/kg of ds)
Fibre sorghum	50–100	25–40	20–30	4,000–4,050
Kenaf	70–100	25–35	10–20	3,700–3,900
Miscanthus	40–70	35–45	15–30	4,200–4,250
Common reed	45–110	35–40	15–35	3,950–4,150
Measly	25–60	35–45	10–25	4,100–4,200
Poplar	20–30	50	10–15	4,100–4,200

So, the biomasses of forest origin and the agricultural residuals, the lignocellulose biomasses from dedicated cultures can be used as fuels in modern plants for heating and more rarely for the combined production of thermal and electrical energy in cogeneration plants. For feeding automatic plants with SRF products and long-term herbaceous cultivations, such as common reed and Miscanthus, it is always preferred to proceed with chipping (par. 4.1.2) of the collected material to make it homogeneous in terms of dimensions. The remaining cultures are more appropriate for briquettes densification (par. 4.1.4) and pellets (par. 4.1.3), but these markets does not exist yet because, at present, wood is the the exclusive raw material for their production.

The choice of the more indicated species must be the result of an attentive evaluation of determinate factors. In fact, although the biomasses of herbaceous origin, coming from long-term cultures, show lower costs of production compared with

Figure 5: Rape, sunflower and soy.

that coming from wooden culture biomasses, a series of obstacles limit their usage in the production of heat and electricity: the lower efficiency during the combustion, the lower specific weight, the lower calorific power for unit weight and the higher cinder content and other undesired compounds such as potassium (K), phosphorus (P), sulphur (S), which are corrosive, or sulphur (S), nitrogen (N) and chlorine (Cl), which are polluting [2, 13].

3.2.2.2 Oil cultures The oil and alcohol-producing cultures, compared to the cultures just discussed, do not directly provide fuel, but instead the raw material from which fuel is obtained by chemical and biochemical transformations.

Many species both arboreal (coconut palm) and herbaceous (sunflower, rape and soy) belong to the oil cultures and they are characterized by a high oil content of the seeds. Sunflower and rape have an intermediate oil content of 48% (with a top of 55%) and 41% (with a top of 50%), respectively. Soy seeds show lower concentrations (18%, with a top of 21%) and they are less appropriate than sunflower and rape for energy production.

The raw oils which are obtained from the oil cultures show an elevated LCP (median 9,400 kcal/kg), so they can be applied as bio-fuels, as a substitute for diesel, for the production of thermal, electric and cogeneration energy. Their conversion into bio-diesel also allows their use for auto traction [1, 2, 7, 14].

Table 5: Oil culture yields in seeds, raw oil and bio-diesel.

Oil culture	Seeds yield (t/ha)	Raw oil yield (t/ha)	Bio-diesel yield (t/ha)
Rape	2.7	1	0.9
Sunflower	3	1.1	1

3.2.2.3 Alcohol-producing cultures The cultures that are named alcohol-producing are those that produce biomass with a high fermentable carbohydrate content that can be applied, through a fermentation process, to the production

Figure 6: Beet, maize and wheat.

of bio-ethanol production. This bio-ethanol can in turn be used as bio-fuel, as a substitute for gasoline or explosion-proof compounds (e.g. methyl *tert*-butly ether (MTBE)).

The raw material that is used at the start of the production line for bio-ethanol can comprise simple sugars, such as sucrose and glucose, or complex sugars (starch), which are obtained, respectively, from dedicated sucrose cultures (sugar beet and sugar sorghum are not the most appropriate for the Italian conditions) or from amylaceous cultures (soft wheat in southern Italy and maize in northern Italy). The simple sugar content of the sucrose cultures is high: the fermentable sugar extract in beet is about 20% of the collected dry biomass, in sorghum it is 18%. The amylaceous cultures contain the starch as grains and the glucose residuals that it is composed of can be hydrolysed and, subsequently, fermented into bio-ethanol: soft wheat has a starch content equal to 70%, for maize it is 78%.

Table 6: Alcohol-producing culture bio-ethanol yields.

Culture	Bio-ethanol yield (t/ha)
Sugar beet	5.5
Sugar sorghum	4.5
Soft wheat	2.5
Maize	3–6

3.3 The zoo technique compartment

The farming wastes produced are termed zootechnical dejections, whereas we speak of dejections only when we refer to the physiological subproducts of the animals (faeces and urine). The composition of the zootechnical dejections varies depending on the origin (cattle/piggish, poultry) and on the farming modality and management. In particular, the water supply (or, on the contrary, the dry substance content) is important for choosing the most appropriate treatment/disposal modalities.

Sewages are the most appropriate for the energy exploitation through anaerobic digestion of zootechnical dejections (par. 2.1, Chapter 4), because they show a dry substance content that is lower than 10–12%.

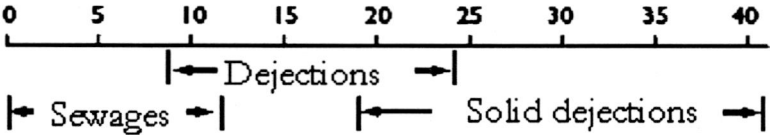

Figure 7: Classification of zootechnical dejections as a function of the dry substance content.

The energy content of the zootechnical sewages is in direct relation with the organic substance content. In fact, it is the organic substance which, through a fermentable or anaerobic digestion process, results in the formation of the bio-gas, a high calorific power fuel.

As evident from the data in reported Table 7, this is the case of the piggish or bovine liquid manures, which are characterized by a high organic substance level (or volatile solids).

Table 7: Bovine and piggish liquid manure yields in bio-gas.

Material	Dry substance (ds; %)	Organic substance (volatile solids – vs; % on the ds)	Bio-gas yield (N m^3/kg vs)
Piggish sewage	3–8	70–80	0.25–0.5
Bovine sewage	5–12	75–85	0.2–0.3

In addition to the quantity of organic substance, it is important to consider the quality of the material; these aspects can, in fact, affect the bio-gas yield and methane content.

The main factors are:

- Composition of the material: It affects the speed of degradation that, in decreasing order, can be schematized as: lignin–cellulose–fats–proteins–carbohydrates. The speed of degradation of a bovine liquid manure, with a higher cellulose material content, is quicker than that of a piggish liquid manure, which is richer in fats (substance that favours higher bio-gas yields).
- Presence of essential elements: Micro-nutrients such as sodium (Na^+), potassium (K^+), calcium (Ca^{2+}), magnesium (Mg^{2+}), ammonia (NH_4^+) and sulphur (S^{2-}), if they are in excess can provoke toxicity. Concentrations even higher than 1 mg/l of heavy metals such as copper (Cu^{2+}), nickel (Ni^{2+}), chromium (Cr^{3+}), zinc (Zn^{2+}) and lead (Pb^{2+}) can be harmful. Other substances that are capable of blocking the digestion are the cleaning and chemical compounds of synthesis [2].

3.4 Industrial activities

3.4.1 The wood industry
In the wood industry three kinds of wastes are produced:

- blank wood wastes (sawdust, small chips, chips);
- treated wood wastes (residuals with glues and/or presence of paints);
- impregnated wood wastes (wood wastes impregnated with salt base preservatives).

Excluding the plants equipped with anti-pollution technology, for energy production purposes, it is possible to only use wood residuals and by-products which are not chemically treated (barking residuals, cut, pruning, etc.) or treated with products that do not contain heavy metals or organic halogenated compounds (typical of wood treated with preservatives or other chemical substances) [2, 15].

The Italian furnishing industry of produces wooden wastes that are equal to 4.7 millions tons a year, of which 55% is not treated wood. Such a considerable residual quantity already has a market: it is applied for energy production purposes or as secondary raw materials for the production of pellets, panels or paper [2, 7, 16].

3.4.2 The cellulose and paper industry
From the paper industry, residuals appropriate for use as raw materials instead of energy residuals are obtained. Such residuals are mainly present as muds and they are generally produced from the water depuration process, both chemical-physical and biological.

In Italy, there is a lack of a strict regulation framework and, as has already happened in other European Community countries, it has resulted in the development of advanced forms of rubbish treatment. In Italy, in fact, only 25% of the energy recovery is obtained from the paper industry residuals against 50% in the European Union. Furthermore, the European Directive 2000/76 does not recognize all the paper production residuals as an adequate or clean fuel and this leads to the obligated disposal in dumps of residuals otherwise usable for the energy recovery [2, 17].

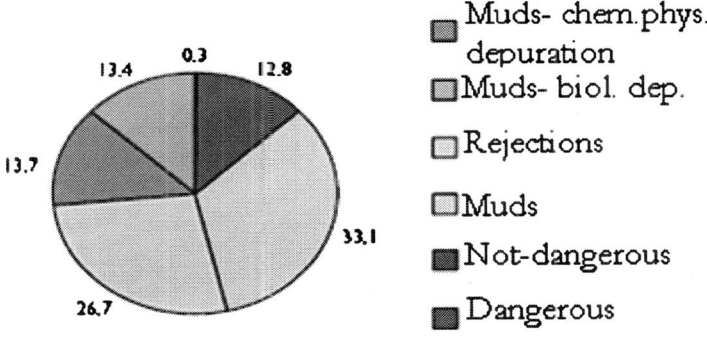

Figure 8: Paper residual typology.

3.4.3 The agro-alimentary industry

Some wastes that are produced from the agro-alimentary industry, because of their organic lead and their high humidity content, are appropriate for the treatment, through anaerobic digestion (par. 2.1, Chapter 4). The main agro-alimentary industry wastes that can be applied to energy recovery through bio-methane, with a specific production of bio-gas between 0.25 and 0.35 m^3/kg ds, are:

- the dairy sector wastes: serum, main cheese working waste with a high organic load; serum can be evaluated from the energy point of view through the anaerobic digestion bio-gas production (in co-digestion with other substrates to avoid an excessive acidification);
- the compartment wastes from butchering (deriving from the meat production effluents for human feed, show high organic loads due to the presence of blood, fat and dung material, in addition to dejections);
- the working and fish conserving wastes;
- beverage industry wastes (particularly the high organic load coming from fruit juice, beer and distillates load wastes);
- the sacchariferous industry wastes (particularly coming from the molasses working which show effluents with a high organic substance content).

3.5 Urban residuals

The urban solid wastes (USW) that, defined as biomass, can be considered as renewable energy sources include all the green biodegradable fractions, which can be divided into those made of lignocellulose wastes component and an humid organic component. The residuals coming from the garden management and public or private boulevards of the habited centres belong to the lignocellulose wastes. Although the urban rubbish is generally used for composing, it can be applied after an appropriate conditioning for heat and/or electricity production through combustion.

Table 8: Product waste composition.

Product fractions of the solid urban wastes	Percentage values
Minus sieve	11.7–12.7
Organic fraction	25.4–29.8
Lignocellulose wastes	3.6–5.8
Paper and cardboards	21.8–24.7
Light plastics	6.9–8.3
Heavy plastics	2.7–3.8
Glass and heavy aggregates	6.7–7.6
Textiles	5.4–6
Metals	2.8–3.5
Leather and Gum	2.4–3.3
Diapers	1.7–2.8

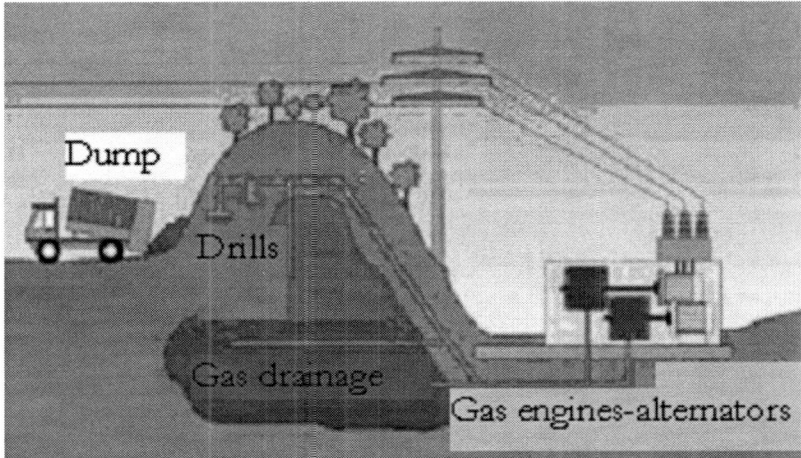

Figure 9: Bio-gas production plant dump frame.

The organic part with a higher humidity content can, however, be applied to the bio-gas through anaerobic digestion production. Fermentation in anaerobic conditions processes also happen in the dumps with a high level of material compaction. In these cases, through appropriate catchment systems (we talk about controlled dumps), it is possible that the cargo and the storage of bio-gas can be applied for energy purposes (Fig. 9) [2].

4 Commercial forms

4.1 Liquid state combustible biomasses

Before being introduced into the market, the lignocellulose biomasses are usually subjected to a transformation process to give them the necessary physical and energy characteristics for their use in the energy plants. Firewood (logs or stub pipes), chips, pellets and briquettes are the main commercial forms for this biomass category.

4.1.1 Firewood
It is sold in logs or stub pipes with variable coal sizes from 50 to 500 mm and humidity values of lower than 50%. This kind of fuel at the domestic level is used mainly in small hand-feed plants, and its use is less than the use of briquettes and pellets (dense forms). The wood boiler, in fact, also does not allow an automatic loading of fuel and shows a lower energy efficiency (50–60% against 75–90% for the chips boilers and wooden pellet) [1, 2].

4.1.2 Chips
To make the wooden and material composition homogeneous and appropriate for the automatic feed in the energy plants, we can resort to the use of

Figure 10: Wooden log and wooden piece boilers.

chips, a mechanical operation that reduces the sortings into flakes of small dimensions that are called chips. Such an operation can be applied, with no difference, to the wooden or herbaceous biomasses. The flakes can be obtained through a crushing-milling process that is essentially based on the percussion and the defibering. The chip is obtained with the cut action.

Figure 11: Wood chips.

The typologies of wood that are prepared for chipping are forest, agricultural and urban pruning residuals, slashes and lops or sawmill subproducts and wood coming from the brief rotation plants (SRF) [1, 2, 20].

The expense during chipping production energy, which is variable depending on the humidity of the biomass used, varies from 2 to 5 kWh/t, which corresponds to

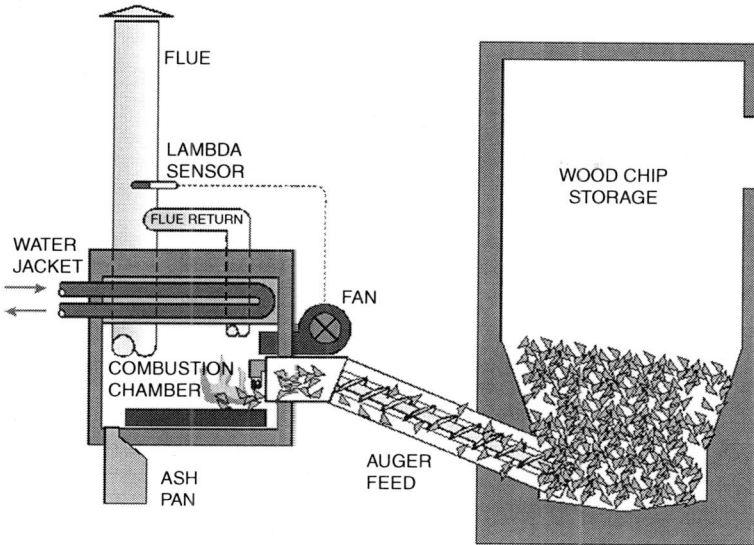

Figure 12: Chip automatic boiler.

less than 0.5% of the energy contained in the wood. Hard and dry wood chipping requires 18% more of energy than the just demolished humid wood working.

The geometry of the chips varies with the cut techniques adopted, as a function of the dimensions required by the type of the energy plant and, especially, of its power system. The chips usually show a variable length from 15 to 50 mm, a width that is equal to half of the length and a variable thickness from 1/5 to 1/10 of the length. A typical dimension is $40 \times 20 \times 3$ mm. The homogeneity (obtained with chip screener adjusting) is the most important parameter for the chips that are destined for combustion, given that non-homogeneous chip dimensions result in annoying blockages of the plant's feed systems.

The standard of the desired humidity is usually obtained prior to cumulus stocking, for an appropriate time: the maximum humidity that is accepted from the disposal of the chips combustion technologies is equal to 50%. It is also important, to avoid chips' working problems, that the standard of the biomass humidity is limited between 25% and 59% [2].

The usable high quality chips in automated combustion systems do not have bark (or it contains only a minimum part) ensuring optimal combustion with a minimum cinder content, which is lower than 0.5%.

Similar to the analogy of the pellet, even for chips it is important to use pure wood. Impurities such as plastic or paints result in higher polluting emissions and cinder content. For this purpose, their use is generally restricted in biomass boilers that do not have a provision for waste gas purification. The chips are good for feeding all the types of biomass boilers with powers that vary from a few kilowatts

Figure 13: On truck set chips.

to tens of megawatts [1, 2]. The market has at its disposal different kinds of chips that are able to provide different quality and dimension wood (up to 30 cm diameter), with a working capacity that varies from a few tons to some tens of tons in an hour, both railcars and carried from agricultural tractors. The cut system can be disks (more diffused and used in the little power chips) or in rolls (that are available in very heavy and powerful versions). For the chips of herbaceous biomasses, cut-waders are often used, machines that work by directly cutting the trunks at the base, chipping them and pushing them on towing [2].

4.1.3 Densified forms: pellets

The birth of the wood pellet can be traced back to 1973 when, in Idaho (USA), an engineer created a new kind of wooden fuel. In the beginning, it was created for an industrial use, but rapidly its use spread to the domestic boilers market. By the term pellet we mean a densified bio-combustible, normally cylindrical-shaped (usually with a variable diameter between 5 and 8 mm and length of 10–20 mm), that is obtained by pressuring the pulverized biomass with or without the help of pressing bidding. At present, the pellet is exclusively obtained from the wooden biomass, but it is necessary to underline the existence of studies that aim to develop the pellet spinneret from herbaceous cultures or from a mixture of these cultures with wooden biomasses. However, the different quality of the starting biomass, especially in terms of calorific power and cinder content, leads to unavoidably different final qualities of the pellet, for different purposes. To avoid these variations, we should make use of the pellet spinneret production technology for the two different biomass typologies [2, 20].

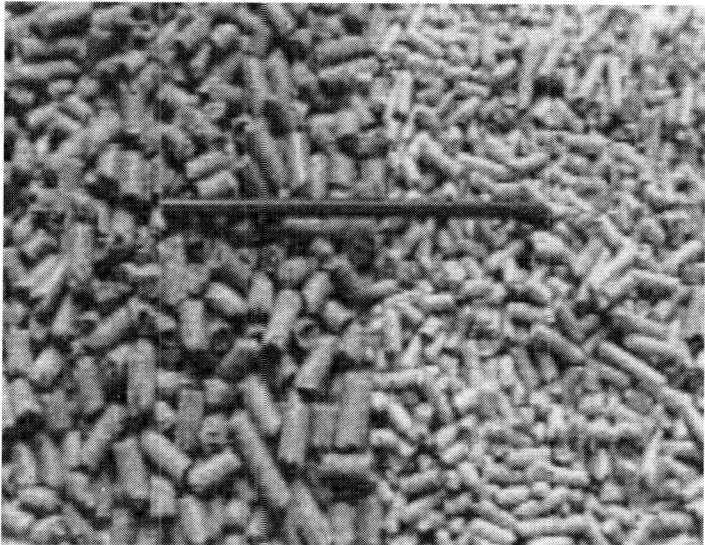

Figure 14: Pellet samples of 10 and 6 mm [10].

Through a frame, the pellet making process can be divided into the following phases: drying, cutting, pelletization, cooling, separation, storage/bagging. In some cases (if the typology of the biomass requires), before the drying phase there are also the phases of roughing, grinding and deferrization. If the working material is rough-shaped (logs or slashes), grinding should be done first, using rotor knives the biomass is reduced to flakes.

Before entering the grinding mill, the raw material is subjected to the magnetic action, which separates from the raw material iron elements (ironing) whose presence could damage the extruder. The most well-known technologies of pelletization do not allow the raw material pressing, if this has a high humidity. After the primary crushing, it is necessary for the material to be dried (usually through a rotary dryer). In this way, the biomass reaches the appropriate humidity grade and allows the lignin that is present in the raw material to be the binding material. At the end of the drying phase, the material reaches a maximum humidity of 10% which allows proceeding with the following treatment steps.

In the second step of grinding, or before the grinding step (if the starting material in the production cycle already has small dimensions: shavings, sawdust, wooden or herbaceous chips, etc.), the material is crushed (usually through a hammer mill) to reduce and level the width down to 3 mm. Such dimensions allow to obtain a standard characteristic for the product. After the material crushing, it passes to the conditioning step, where it is prepared to enter the pelletization spinneret. This phase can also include the embedding of bidding or additive agents. Usually, this conditioning operation is realized using dry water vapour to frizz the wooden fibres and to induce a partial gelling of the biomass.

The pelletization that results from compression comprises perforated, cylindrical and flat forms (matrix) through whose holes the conditioned biomass is pushed at high pressure (up to 200 atm) using roll systems. The pellet is formed due to the transformations that happen during the passage of the fibre through the extrusion holes, when the temperature is up to 90°C. At these temperatures, there is a fluidization of the lignin that comes out from the cellular structure; this allows the fibres to stick to each other. Appropriate blades cut, at the desired length, the compressed material and the surface bakelized that overflows from the matrix holes. The extruded and cut material then passes to the following cooling step (realized through ventilation plants), where the product undergoes hardening. Subsequently, in the separation section, the whole pellet is deleted and it is re-entered in the extrusion system. At the end of the cycle, the pellet is stored in silos or bagged (storage/bagging steps) [2]. The mechanical energy spent for the pellet production is equal to the 2% of the final product's energy content. As a consequence, pellets are considerably better than the fossil energy sources, for which 10–12% of their energy content is used for refining [1]. The cost of pellet production varies between 0.05 and 0.16 €/kg based on the applied technology and on the biomass used. The wholesale price is equal to 0.11–0.21 €/kg, whereas the detail price is 0.21–0.30 €/kg [20]. The national pellet production, which was estimated at around 240,000 tons in the 2004/2005 season, shows a strong growing trend [22].

The pellet changing allows obtaining a product characterized by high energy density. It is also easily transportable: in terms of motion, in fact, it acts similar to fluids. This allows a high grade of automation of the tools and the combustion plants. This pellet property is due to the particular from, dimension and homogeneity of its elements that can be sent to the combustion oven through simple mechanical instruments (transport foils, scrolls or suction systems), with important advantages such as automatic control, dosing and continuous feed [1, 2].

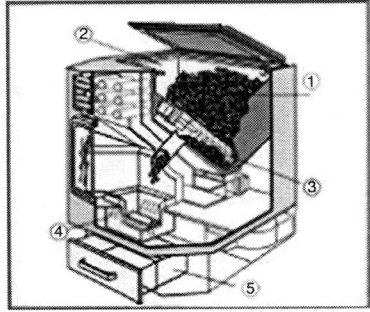
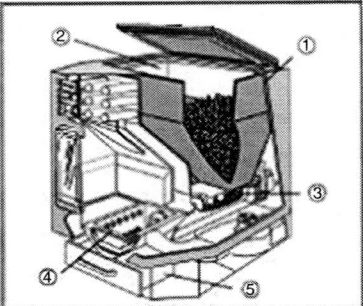

Figure 15: Pellet boiler frame. 1: pellet; 2: tank; 3: fuel pipe; 4: boiler furnace; 5: ashes.

Compared to the non-densified biomasses (sawdust, chips and slashes, etc.), pellets show some advantages that make it more valuable in the market [2]:

1. High apparent density (bulk density, bio-fuel volume for unit mass): variable between 650 and 780 kg/m^3. The pellet density is seven times higher than the density of sawdust and chips and this optimizes its transport and storage.
2. Low humidity content: low hydric content improves the combustion yield and contributes to reduced transportation costs. Furthermore, during the storage phase, the combustible does not show risk of undergoing fermentative phenomena.
3. High calorific power for unit weight: it depends on the composition and the structure of the biomass applied for pellet making. The wood pellet has a LCP of about 4,000 kcal/kg (high energetic weight among the bio-fuels).
4. Homogeneity of the material has both physical and qualitative characteristics points of view: The first, together with the small dimensions of the pellet, allows to easily move the product through scrolls, transport foils or wheel suction systems and the possibility to use them in automotive boilers, whereas the second allows a better regulation of the combustion and a better control of the emissions.

Table 9: Energetic equivalence between pellet and fossil fuels [2].

	Pellet (kg)	Diesel (l)	Diesel (kg)	Combustible oil (kg)	Natural gas (m^3)	GPL (kg)	Coal (kg)
1 kg of pellet	1	0.46	0.4	0.41	0.49	0.37	0.58
1 l of diesel	2.17	1	0.86	0.89	1.06	0.8	1.25
1 kg of diesel	2.52	1.16	1	1.04	1.24	0.93	1.46
1 kg of combustible oil	2.43	1.12	0.96	1	1.19	0.89	1.4
1 m^3 of natural gas	2.04	0.94	0.81	0.84	1	0.75	1.18
1 kg of GPL	2.72	1.26	1.08	1.12	1.33	1	1.57
1 kg of coal	1.73	0.8	0.69	0.71	0.85	0.64	1
LCP (kW h/qtity)	4.70	10.19	11.86	11.40	9.59	12.79	8.14

4.1.4 Densified forms: the briquette

Similar to pelletization, the briquette also represents a particularly interesting technology because by reducing the material density it allows to concentrate high energy reserves in a contained volume. The briquette, in fact, is a bio-fuel that has a parallelepiped or cylindrical shape; it is obtained by compressing some pulverized biomass with or without the help of pressing additives. During the process of production, the wood is desiccated so that humidity is not higher than 10%. Briquettes can be stored easily and their calorific power is 18.5 MJ/kg [1, 2].

Figure 16: Briquettes.

A complete line of briquetting is usually composed of subsequent steps that change the raw biomass, with variable humidity and pressing characteristics, into a standardized briquette that is ready to be sold in the market.

The line comprises steps that are very similar to those for pelletization, but with a relatively simpler technology. In principle, there is a biomass pre-treatment which is followed by compaction and by briquette changing.

Crushing, drying and biomass heating, which are necessary to reach the optimal characteristics of granulometry and water content, form the pre-treatment phase.

Through a feed system, which comprises holding under pressure, or by a conveyor belt pipe, the pre-treated biomass is brought to the briquette phase, where it is compacted and transformed into briquettes. We can have low-, medium- and high-pressure briquetting systems depending on the pressure exerted. In the first two cases, the biomass is mixed with a binding substance, whereas the high-pressure systems work on the biomass as well as the biding effect obtained after the high pressures are exerted.

The most important high-pressure briquette technologies are screw briquetting and the piston briquetting (mechanic oleodynamic). In screw briquetting, the biomass is continuously extruded because of the rotation of one or more screws without the end inside a cone room (heated during the process). During mechanic piston briquetting, the compaction of the biomass is achieved through an alternative piston which is actuated by an electrical engine, whereas during the oleodynamic circuit piston process, there are two pistons which are actuated by the holding pressure of a closed-circuit oil which compress the material in orthogonal directions.

The most used system, which allows the treatment of the biomass with higher humidity and a better control of the applied pressure, is piston briquetting.

The final transformation of the product which is obtained from the briquetting phase includes the steps of the eventual briquette cut (only for screw systems where the material continuously exits), eventual cooling (for screw systems with heating), packing and final stocking of the briquettes [2].

Table 10: Comparison between piston press and screw press [2].

Characteristic	Piston press	Screw press
Optimal humidity biomass (%)	10–15	8–9
Fatigue of the mechanic parts (–)	Low	High
Briquetted product	In blocks	Continuously
Energy medium consumption (kW h/t)	50	60
Briquette specific weight (kg/m^3)	1,000–1,200	1,000–1,400
Combustion briquette behaviour	Medium	Very good
External carbonization	Absent	Present
Briquette homogeneity	Scarce	Good

The densification of the biomass in briquettes has the same advantages and disadvantages as the transformation into pellets: in fact, after the briquetting process we obtain an improvement in the physical biomass characteristics (density, homogeneity, etc.), a reduction in the volume, a reduction in the storage and transport costs, and an improvement in the behaviour during combustion. At the same time, the briquetting process, just as the pelletization process, needs, as we have seen, a preventive material conditioning, in particular the biomass drying. The briquettes can be used in the place of firewood and coal, by adjusting some operative parameters such as the primary and secondary air distribution; relative to the two fuels, in fact, the briquettes require a higher quantity of secondary air and a lower quantity of primary air.

The briquettes with a higher thermal capacity (retain heat for a longer period of time and keep the temperature inside the oven high to allow easy combustion of the newly entering combustible) are a 'better' combustible than uncompressed wood.

This type of combustible is used most frequently in both domestic (they do not generate sparks which makes them appropriate for the chimney) and industrial applications [1, 2].

4.1.5 Lignocellulose biomass fitness for the transformation into commercial forms

The several kinds of lignocellulose biomass (arboreal, perennial herbaceous, annual herbaceous) that have different applications. The wooden biomasses are applied to the selling of logs or wood trunks, as well as to chipping and pellet or briquette transformation.

In the first case, the use is limited to small domestic heating plants, whereas chipping is necessary for the use of the biomass in large heat generation industrial plants. The briquettes optimally substitute the firewood. The low quality pellet, not suitable for the feed in small plants, can be efficiently used in industrial plants [2, 13].

Herbaceous biomasses are appropriate for the production of chips or pellets which, because of the lower quality, are not suitable for use in small plants, but they find better application in large dimensions equipped with appropriate expedients to optimize combustion plants. A possible future scenario is the mixing of the herbaceous and wooden biomass to exploit, in a synergic way, the peculiarities of the two combustibles [2].

4.2 Fuel biomass in the liquid state

4.2.1 Vegetable oils

From sunflower seeds and rape seeds, oils are extracted which can, actually, be used without the need of further bio-fuels; therefore, they can be applied to the energy range just as combustibles of liquid fossil origin. Table 11 lists the main energy characteristics of combustible oils and compares them with the characteristics of gasoline.

Table 11: Comparison between the properties of vegetable combustible oil fuels and diesel [2].

Parameters	Measure unit	Vegetable combustible oils	Diesel
LCP	kcal/kg	9,000–9,500	10,200
Flashpoint	°C	230–290	60
Cetane number	–	30–40	54
Density	kg/m^3	0.915	0.839
Viscosity at 38°C	mm^2/s	27–53	2.7

The current energy applications of vegetable oils are relative to the use in diesel engines. Their use in energy production plants as a substitute for diesel poses some problems: the burners must be partially changed to cover the higher viscosity of the vegetable oils compared to diesel.

The high viscosity also excludes, at present, the application of vegetable oils for auto traction. In fact, in this case important changes would be necessary to the plans of engines [2, 14]. Vegetable oils are composed of 78% carbon, 12% hydrogen and 10% oxygen.

Oils are organic compounds, so they are highly biodegradable; these characteristics can represent an obstacle in their use as a combustible because they can undergo oxidation and polymerization in the storage reservoirs. For this reason, oils must be applied within 12 months from their production. Keeping in mind that the viscosity of the oils gradually increases as the temperature decreases (until solidification), such combustibles are not appropriate for use at temperatures below 5°C [1].

4.2.2 Bio-diesel

The limitations posed by the use of vegetable oils in some typologies engines, in particular those that require auto traction, due to the high viscosity of the combustible, can be overcome by the transformation of oils into bio-diesel, which is obtained by the esterification process (this will be analysed in detail in the par. 2.4.3, Chapter 4). Vegetable oils are mainly made of triglycerides and their viscosity (median of 40 mm^2/s) is higher than that of fuels of fossil origin (diesel has a viscosity equal to 3 mm^2/s). The transformation into bio-diesel is due to the conversion of the triglycerides into methylic esters which reduces the product's viscosity. The mixture of methylic esters, called bio-diesel, has a viscosity of 5 mm^2/s which is similar to that of diesel. Bio-diesel can be produced even from saturated cooking oils. These substances must be purified before the esterification process. The esterification process involves the use of alcohol, mainly methanol, and it yields glycerine as a by-product (which is re-sold to the chemical and pharmaceutical industry as a raw material) [1, 2]. Consequently, by using bio-diesel, on one hand, an improvement in the invested capitals is achieved (compared to that necessary to start the spinneret which aims to produce energy using vegetable oils); on the other hand, there is a noticeable increase in the possible use scenarios. From Table 12 we can see that bio-diesel shows some characteristics which are similar to diesel.

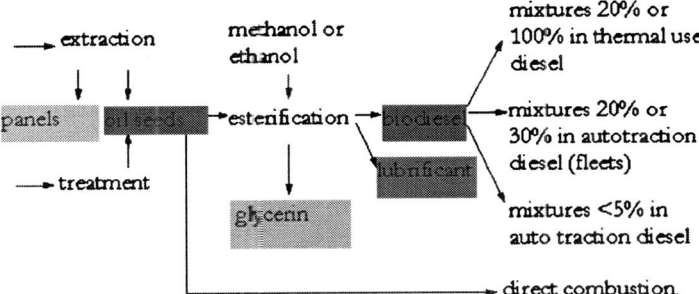

Figure 17: Bio-diesel spinneret.

Table 12: Comparison between the properties of bio-diesel and diesel fuels [2].

Parameters	Measure unit	Bio-diesel	Diesel
LCP	kcal/kg	8,900	10,200
Flashpoint	°C	85–178	63
Viscosity at 38°C	mm^2/s	4.78	3.12
Cetane number	–	48–56	54
Density	kg/m^3	885	839

The substitution of diesel for auto traction in vehicles equipped with diesel engines and the feeding of boilers for electricity generation are the main uses of bio-diesel as an energy product. The recent interest shown by the European Committee (Directive 2003/30 of the European Parliament and of the Council, 8 May 2003) in the development of the biocombustible spinneret and in view of the fact that it is already theoretically possible to feed diesel engines which are present in the market with a mixture of diesel and bio-diesel make the use of bio-diesel as a fuel for auto traction particularly interesting.

Directive 2003/30 of the European Parliament and of the Council, 8 May 2003, on the promotion of bio-fuels and other renewable fuels in the transport sector, stipulates that each member state should fix its target for bio-fuel usage relative to the bio-fuel quotation available in the market. Such targets must be based on the reference levels as per the Directive: 2% of the total fuels (petrol and diesel) that are available in market by 31 December 2005 and 5.75% by 31 December 2010. Legislative Decree 120/2005 acknowledges Directive 2003/30 at a national level and for Italy it establishes the target of 1% by 31 December 2005 and 2.5% by 31 December 2010.

Currently, in Italy, the use of bio-diesel for auto traction is limited to the realization of mixtures with diesel only up to 5%. Studies conducted by German and Austrian experts demonstrated the possibility of using bio-diesel in mixtures up to 30% without making any modifications in the engine. To use pure bio-diesel, on the contrary, it is necessary to substitute the rubber rings with other compatible materials (copper, carbon steel, brass, fluorinated rubbers, etc.).

Bio-diesel, pure or as a mixture, can also be used in diesel burners by making only modest remedial interventions (regulation of the air/fuel ratio, modification of the atomization nozzles' slant, etc.) [2, 14, 24].

The main advantages of bio-diesel compared to the use of the traditional diesel are [24]

- high number of cetanes (higher flammability in diesel cycles);
- high lubricant capacity;
- absence of sulphur;
- high percentage of oxygen (high stability of combustion, lower production of PM10, lower volatile organic residuals);

It is useful to remember that bio-diesel production is an option for the agricultural cultivation of the abandoned or rotated soils after intensive cultures. Figure 18 shows the historical trends of bio-diesel production in Europe.

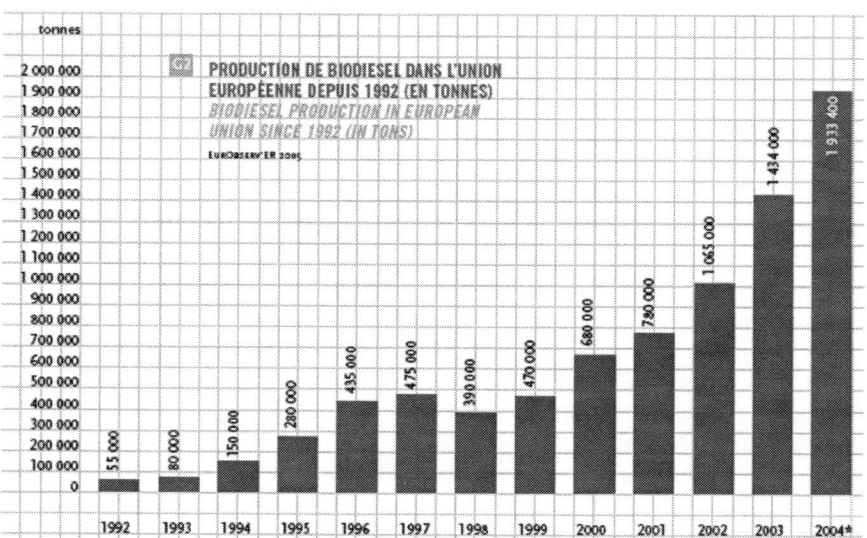

Figure 18:   Historical trends of bio-diesel production in Europe [25].

The statistical almanac on bio-diesel production, edited by the European Bio-diesel Board (EBB), describes the highly positive bio-fuel production situation in 2005. Particularly in Europe, an increase in production equal to 65% of the production in the previous year was observed. This increment was a result of the involvement of 15 member states of the European Community, of whom Germany again affirmed itself as the leader in bio-diesel production with 1.6 millions tons produced in 2005. Although the scenario is positive, the target of 2% of the biocombustible production as per European Directive 2003/30 has not yet been achieved [26].

Table 13: Bio-diesel production and capacity of production in Europe (thousands of tons) [27].

Nation	2004 production	2005 production	2005 capacity	2006 capacity
Germany	1035	1699	1903	2681
France	348	492	532	775
Italy	320	396	827	857
Czech Republic	60	133	188	203
Poland	0	100	100	150
Austria	57	85	125	134
Slovakia	15	78	89	89
Spain	13	73	100	244
Denmark	70	71	81	81
UK	9	51	129	445
Slovenia	0	8	17	17
Estonia	0	7	10	20
Lithuania	5	7	10	10
Latvia	0	5	5	8
Greece	0	3	35	75
Malta	0	2	2	3
Belgium	0	1	55	85
Cyprus	0	1	2	2
Portugal	0	1	6	146
Sweden	1.4	1	12	52
Total	1933.4	3184	4228	6069

Figure 20 shows the trends for bio-diesel spinneret development Italy.

4.2.3 Bio-ethanol

Ethanol can be produced chemically starting from a fossil source or by fermentation starting from biomasses. This second method leads to the so-called bio-ethanol production.

Bio-ethanol is a vegetable origin fuel which is obtained by the fermentation of alcohol from sugars and complex carbohydrates, such as starch, cellulose and hemicellulose. However, the raw materials for ethanol production can be derived from alcohol-producing dedicated cultures if they are sacchariferous (sugar beet, sugar

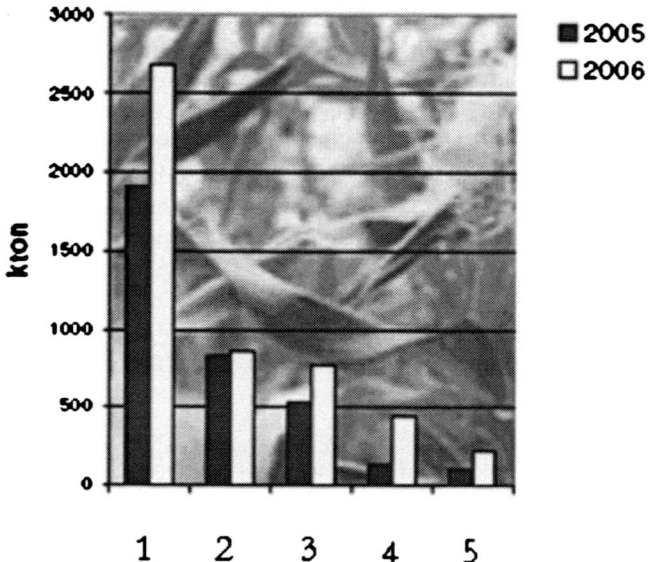

Figure 19: Bio-diesel production capacity in 2005 and 2006 in some of the EU countries. 1: Germany; 2: Italy; 3: France; 4: United Kingdom; 5: Spain.

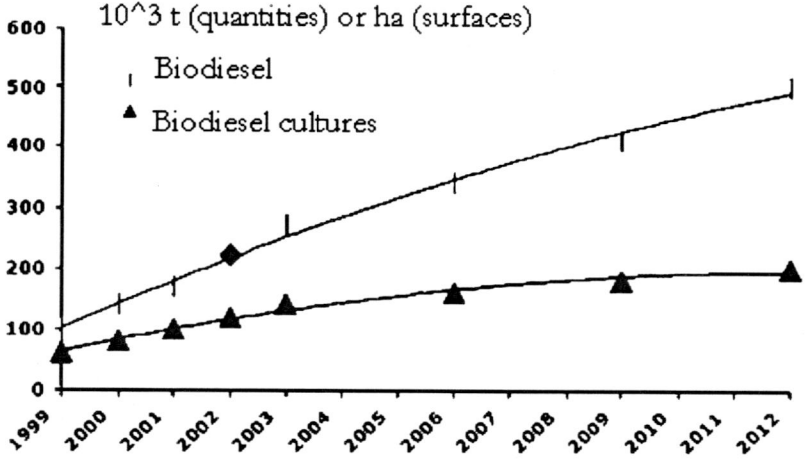

Figure 20: Bio-diesel spinneret development trends in Italy.

cane, sugar sorghum) or starchy (soft wheat and corn) as well as from lignocellulosic residuals obtained from the forest and agricultural workings. The last materials mentioned do not require specific workings, as in the case of dedicated cultures, which reduces the cost for the retrieval of the raw material; therefore, they represent the most interesting option from the economical point of view.

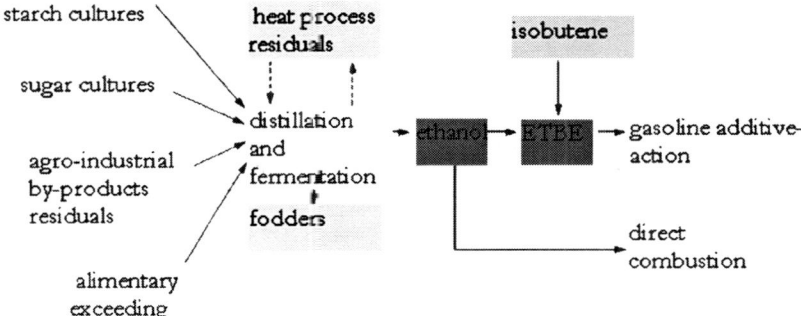

Figure 21: Bio-ethanol spinneret.

The bio-ethanol production spinneret comprises three sections (analysed in detail in par. 2.3, Chapter 4): sacchariferous, starchy and lignocellulose.

In addition to the biomasses described above, as raw materials for bio-ethanol, agricultural and food industries residuals and urban wastes can also be used, and based on their nature they can be included in one of the three categories of the bio-combustible production spinneret.

Bio-ethanol can be mixed with gasoline or, in some cases using appropriate expedients, it can be substituted as the feed in vehicles; this bio-combustible, in fact, shows chemical–physical characteristics that are similar to gasoline. Table 14 lists the main energy characteristics of bio-ethanol compared with those of gasoline [1, 2, 24, 29].

Table 14: Comparison between the properties of bio-ethanol and gasoline fuels [2].

Parameters	Measure unit	Bio-ethanol	Gasoline
LCP	kcal/kg	6,500	10,500
Flashpoint	°C	13	21
Boiling temperature	°C to 1 atm	78	105

The country that stands out for the use of bio-ethanol is Brazil where, even in the 1970s, the engines were modified for the use of anhydrous bio-ethanol (with 5% of water residuals as a substitute for gasoline. This practice is today a reality in the South American country. The interventions needed to adapt engines for the use of anhydrous bio-ethanol as a substitute for gasoline include valves regulation and replacement of the components that can corrode. In USA and Canada, on the contrary, anhydrous bio-ethanol is used in mixtures with gasoline up to 10% in non-modified engines and up to 85% in modified engines. The feeding of these latter engines, called flexible fuel vehicles (FVV), can be realized either with gasoline and bio-ethanol mixtures or with gasoline only; in fact, they are equipped with the automatic regulation of the injection times and the ratio of air–fuel mixing. In European and American studies, the possibility of using bio-ethanol in mixtures up to 23.5% without changes to the motor has been emphasized.

At present, in Europe, the presence of anhydrous bio-ethanol in gasoline in concentrations up to 5% is allowed. It is also important to underline that the properties of bio-ethanol increases the engine efficiency and reduces the fuel combustion [1, 2, 24].

Instead of bio-ethanol it is possible to use ethyl *tert*-butly ether (ETBE), obtained by the reaction between ethanol and isobutylene in the presence of appropriate catalysts, which finds use as an antiknock with a high octane number. ETBE can be used in place of benzene (carcinogen) and methyl *tert*-butyl ether (MTBE; highly polluting, especially in subterranean waters) compared to which it has a lower environmental and human health impact. Furthermore, ETBE, if it is used in mixtures with gasoline at 15% gives an octane number equal to 110, which is higher than the octane number of 95–98 that is typical of traditional antiknocks [2, 30]. Some studies have also demonstrated the possibility of using bio-ethanol in mixtures with diesel: up to mixtures of 15%, without any modification to the diesel engine.

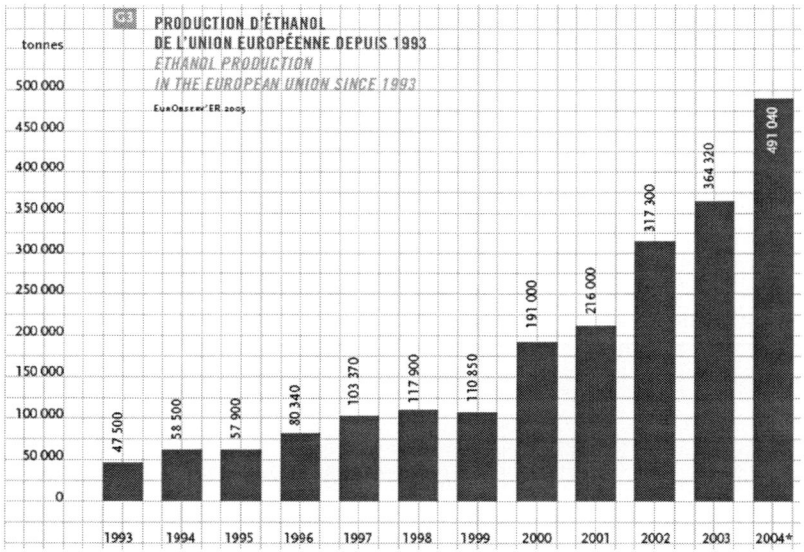

Figure 22: Historical trends of bio-ethanol production in Europe [27].

4.3 Combustible biomasses in the gaseous state

4.3.1 Bio-gas

Bio-gas is a combustible with high calorific power (4,500–6,500 kcal/N m^3 depending on the chemical composition of the gas) that is obtained by anaerobic digestion (see par. 2.1, Chapter 4) of an organic substance. The main components of bio-gas are methane (CH_4) and carbon dioxide (CO_2); other substances with a lesser percentage are carbon monoxide (CO), nitrogen (N_2), hydrogen (H_2) and hydrogen sulphide (H_2S).

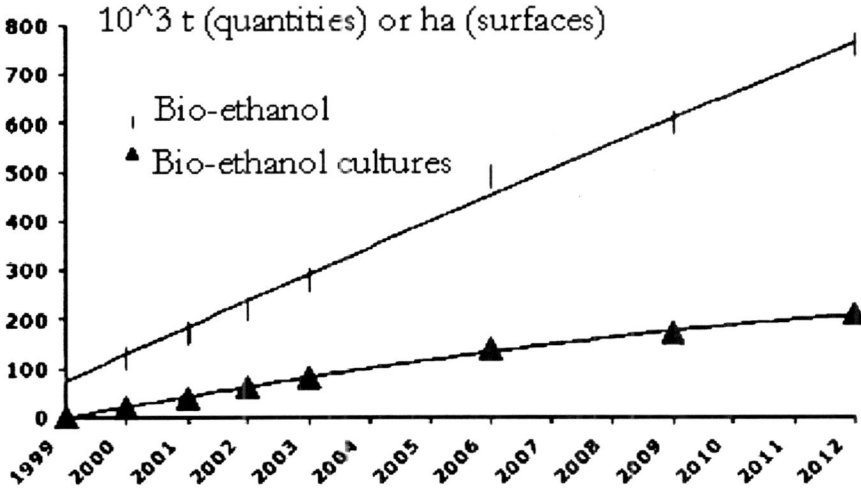

Figure 23: Predictions for bio-ethanol spinneret development in Italy.

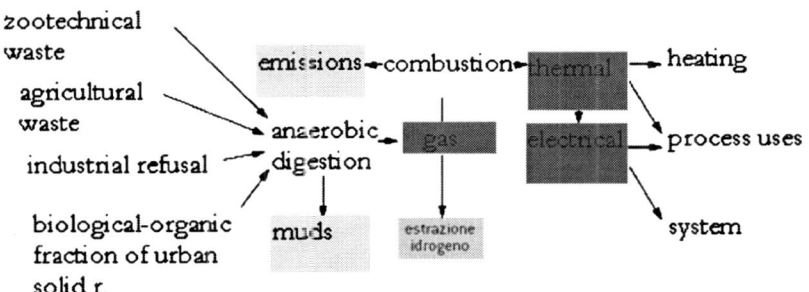

Figure 24: Bio-gas spinneret.

Table 15: Medium gas composition [2].

Compounds	% (on dry gas)
Methane (CH_4)	50–80
Carbon dioxide (CO_2)	35–45
Hydrogen sulphide (H_2S)	0.02–0.2
Water vapour	Saturation
Hydrogen, ammonia	Traces
Oxygen, nitrogen	Traces

Before being used for energy production, the bio-gas must be subjected to appropriate treatments (see Table 17) which, by raising the percentage of methane to the detriment of other gases, improves the calorific power.

Table 16: Energy equivalence between 1 m³ of bio-gas and the main combustibles.

Gasoline	0.8 l
Methane	0.7 m³
Ethylic alcohol	1.3 l
Wood coal	1.4 kg
Wood	2.7 kg

In fact, it is the concentration of methane in the mixture that determines the final calorific power of the gas: the higher the concentration of methane, the higher is the LCP; the presence of carbon dioxide, nitrogen and water has a contrary effect. The treatment to which the bio-gas is subjected should reduce the percentage of corrosive agents, such as hydrogen sulphide, which can damage the utilization plants. The choice of the treatment, or treatments, to which the bio-gas is subjected depends on its initial characteristics and on the final calculated utilization [1, 2, 31].

Table 17: Bio-gas treatments (*only in the presence of excessive H_2S).

	Consequence	Treatment	Usage
Water	Condense that provokes malfunctioning Corrosive potential action	Condense separators Refrigerator condensation systems	In boiler Cogeneration ≈ natural gas
H_2S	Engine corrosion Electrical elements	Iron oxides filters Active coal filters Biofilters NaOH washing Iron salts solution washing	In boiler* Cogeneration* ≈ natural gas
CO_2	Removal of CO_2 is necessary to improve the level of methane in the bio-gas (≈ natural gas)	Absorbing in water with subsequent stripping and emission in atmosphere Half-permeable Membranes that selectively retain CH_4	≈ Natural gas

Table 18: Biomasses and anaerobic digestion organic wastes and their bio-gas yield (m³ per volatile solid ton) [32].

Materials	m³ bio-gas/t SV
Animal dejections (piggish, bovine, birds and rabbit)	200–500
Cultural residuals (straw, beet collar, etc.)	350–400
Agroindustrial organic wastes (serum, vegetable wastes, yeast, muds, distillery effluents, beerhouse, cellar, etc.)	400–800
Abattoir wastes (fats, stomach and intestinal content, flotation sludge, etc.)	550–1,000
Depuration muds	250–350
Urban wastes organic fraction	400–600
Energy cultures (maize, sugar sorghum, etc.)	550–750

Currently, the main uses of bio-gas are relative to the thermal and/or electrical energy production. In detail, it is possible to produce [2, 32]:

- heat, as hot water, vapour or air, with a medium energy efficiency of 80–85%;
- electricity, generally in engines with vapour or gas turbines, for plants with a high capacity whose medium yield is 30–35%;
- combined production of heat and electricity (cogeneration) in endothermic engines with total medium yields of 80–85% (medium thermal yield: 50%, medium electric yield: 35%), which is currently the most used solution.

Other emerging applications are [2, 32]:

- fuel production for vehicles;
- cold production (three-generation), for example, with absorbing machines;
- the use in industrial ovens as a primary or auxiliary combustible.

In Europe, since 1990 we have witnessed a continuous growth in bio-gas production from 2,304 tons in 1999 to 3,219 tons in 2003. The leading country in this sector continues to be England, with more than 15,000 technicians in the sector in 2003 and a production of raw bio-gas which is equal to 1,151 tons; then there is Germany which has declared 2.000 installations for bio-gas production that is equal to 685 tons [33].

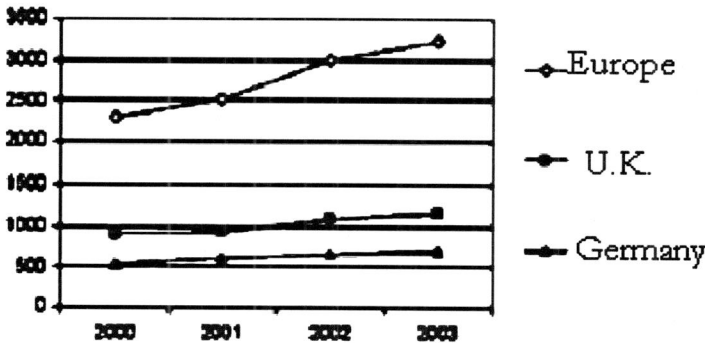

Figure 25: Bio-gas production based on the geographical area ($t \times 1,000$).

Currently, Europe can count on [32]:

- 1,600 operative digestion tanks for the stabilization of depuration sludge;
- 400 bio-gas plants for the effluent industrial waters with high organic load;
- 450 plants that work on the recovery of bio-gas from the urban rubbish dumps;

- more than 2,500 plants that work on effluents from intensive animal breeding, particularly in Germany (>2,000), Italy, Austria, Denmark and Sweden;
- 130 plants for anaerobic digestion; each of them processes more than 2,500 tons in a year of urban rejections and or industrial organic residuals with organic content.

In Italy, on the contrary, there are [32]:

- 120 digestion tanks for the stabilization of sludge and purification of effluent urban waters;
- few experiences (seven plants) of anaerobic digestion of urban rejections with organic content;
- several bio-gas plants in the agro-industry;
- more than 100 bio-gas plants for effluents from intensive animal breeding.

CHAPTER 4

Energy from biomasses

1 Biomass energy conversion

There are multiple methods for the conversion of biomass energy. Generally, the biomass is transformed into a more easily manageable form (solid, liquid or gaseous bio-fuel) in which it is used [1, 3, 5]. Before proceeding with the analysis of the main modalities of biomass energy conversion, it is important to note that conversion is only one aspect of the wider problem, which includes, on one hand, the present reality or the eventual future situation where the biomasses are produced and, on the other hand, the possible utilization of the deliverable energies. The resulting circuit is depicted as follows:

Production–Collection–Conversion–Utilization

This has to be studied in an optimal context that includes extensive co-ordinated initiatives and interventions with both public and the entrepreneurial support. The biomass energy conversion processes are divided into biochemical and thermochemical processes [3–5, 34, 35]:

- Biochemical conversion processes allow to exploit the energy obtained through chemical reactions, due to the action of enzymes, fungi and micro-organisms, that take place in the biomass under particular conditions. These processes are suitable for those biomasses in which the C/N ratio is lower than 30 and the humidity at the time of collection is higher than 30%. Water cultivations that are appropriate for biochemical conversion include the by-products of some cultures (leaves and beet stems, vegetable garden, potato, etc.), livestock sewage wastes and some working wastes (pot ale, vegetation water, etc.) as well as the heterogeneous stored biomasses in the controlled biomasses.
- Thermochemical processes are based on the action of heat which allows the chemical reactions that are necessary to transform the material into energy to take place. These processes are suitable for products such as cellulose and wooden residuals in which the C/N ratio has values higher than 30 and the humidity content does go beyond 30%. The most appropriate biomasses for

thermochemical conversion processes are wood and all its by-products (sawdust, shavings, etc.), the most common cultural by-products of the lignocellulosic kind (straw of cereals and orchards, vine pruning residuals, etc.) and working wastes (such as husk, chaff, hulls, hazels).

Of the several technologies available for energy conversion of biomasses, some of them can be considered to be at the development level to allow their utilization on an industrial scale, while others, on the contrary, need further experimentation to raise their efficiencies and reduce the energy conversion costs.

Usually, upstream, all the conversion processes require appropriate base material for pre-treatment, which can include water washing, drying with mechanical or thermal instruments (pressing), reduction into smaller dimensions, densification (pelletization and briquetting) and separation of the fibres (extraction with solvents). Even the final product, depending on its use, must be treated to separate them (e.g. from the substrate that does not react, catalysts, micro-organisms and solvents), purify them and concentrate them. Depending on the application, we resort to sedimentation, filtration, centrifugation, distillation, absorption and extraction with solvents [4].

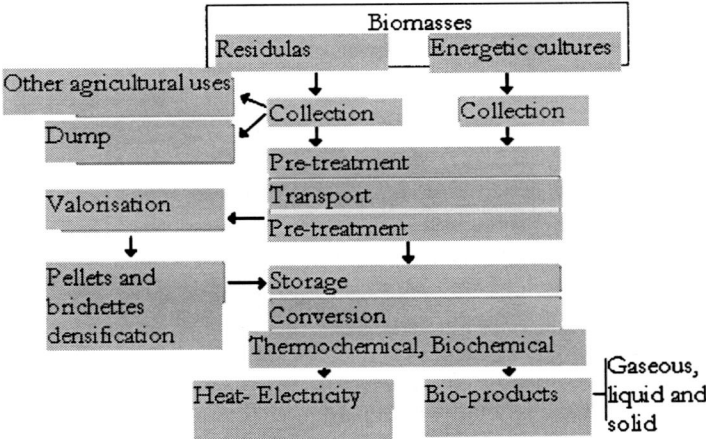

Figure 1: Biomass spinneret from collection to transformation.

2 Biochemical conversion

The chemical conversion processes include [3, 36]:

- anaerobic digestion;
- aerobic digestion;
- alcohol fermentation;
- extraction from oils and bio-diesel production.

2.1 Anaerobic digestion

The fermentation of methane, also known as anaerobic digestion, is a complex natural process that involves bio-degradation of the organic substance in the absence of oxygen (anaerobiosis), resulting in the formation of bio-gas.

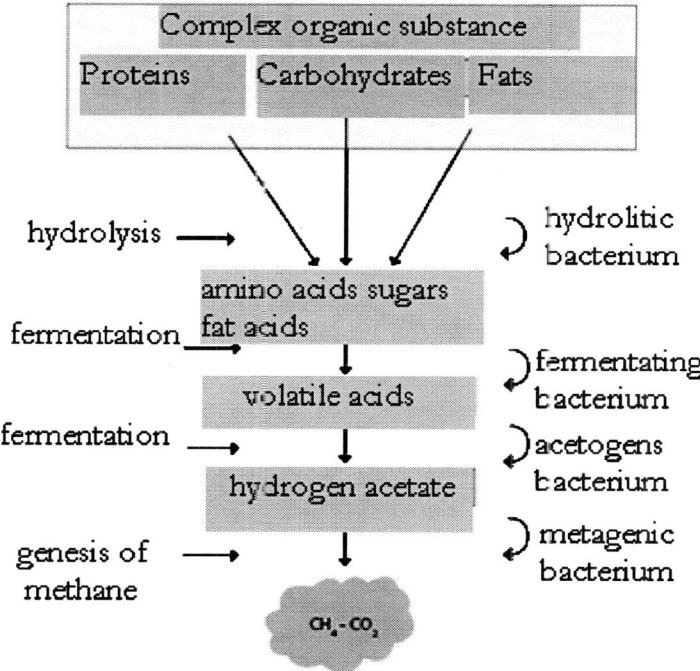

Figure 2: Anaerobic digestion process steps.

As we can see from Fig. 2, the anaerobic digestion process comprises three sequential steps involving different bacterial groups that act in series. In the first step (hydrolysis), the hydrolytic bacterium breaks the complex organic compounds (carbohydrates, proteins and fats) into simpler substances. Subsequently (in the fermentation step), these simpler substances are transformed, first, into organic acids, through acid-genesis reactions, and then into acetate, carbon dioxide and hydrogen, through vinegar-genesis processes. In the last step, the most important step (methane genesis), the methanogenic bacterium transforms the products that are formed in the previous step into methane (CH_4) and carbon dioxide (CO_2), the main constituents of bio-gas. Therefore, the organic component is degraded releasing the chemical energy it contains in the form of bio-gas.

As evident from the above description, bio-gas production depends on the coordinated and sequential action of all the microbic groups involved. To achieve this goal, it is essential that the reaction environment is the result of a compromise between the requirements of each individual group involved, by a strict control of the process parameters [2, 31]. The anaerobic digestion processes can be classified based on the mass fraction of the dry substance to be digested: if it is lower than 10%, the process is called wet digestion; if it is between 10% and 20%, the process is called semi-dry digestion, and for values higher than 20%, it is called dry digestion [31, 32].

Depending on the temperature range in which the process takes place, anaerobic digestion is called:

- psychrophilous: if the process temperature is kept below 20°C; the systems that work in such conditions are also called 'cold';
- mesophily: if the temperature is between 20°C and 40°C;
- thermophile: if the temperature of the process is between 50°C and 65°C.

The processes described above can be reproduced in confined environments such as an anaerobic methane digester (for the digestion of liquid manures with a high organic load) or in a controlled dump (for the digestion of the organic component of solid rejections) [2, 31, 32].

2.1.1 Plant typologies applicable to liquid or effluent manures

The engineering typologies of the anaerobic digestion systems that are currently available vary from extremely simple systems that are mainly applied to the livestock sewage wastes on a business scale to those that are more sophisticated and use high technology and are applied for industrial effluent treatment.

2.1.1.1 Simplified plants Due to its constructive and managerial simplicity, this plant typology finds many possibilities for application in the zoo technical sector. These plants, in fact, only comprise a storage basin (containing the material to be digested), often pre-existing, equipped with an appropriate gasometric cover. The simpler systems are the 'cold' systems (psychrophiles) that have variable yields depending on the season and elevated permanence times (around 60 days). The annual bio-gas production for a swine liquid manure is about 25 m^3/100 kg live weight. The systems that are equipped with heating, on the contrary, obtained from the bio-gas produced, allow to work in a mesophily regime and to obtain higher, more constant yields during the year, with more reduced retention times (median of 20 days). In this case, the annual production of bio-gas from swine liquid manure will be around 32 m^3/100 kg live weight [2, 32].

The function of the gasometric cover is to retain and store the bio-gas that is formed; it can be dome shaped or have a floating shape [2]:

- Simple dome cover: It is not pressurized and is made of flexible canvas material that is anchored on the basin's perimeter. The gas, being at very low pressure, is extracted and sent to its place of use through a blower.

- Double or triple membrane dome cover: This type of cover comprises two or three superimposed membrane layers which are fixed at the edge of the basin (see Fig. 3). In this case, the draining of the bio-gas is achieved by overpressure valves that are regulated by sensors.
- Floating cover: These are membranes which are equipped with a ballast system that is realized using flexible pipes filled with water to grant the bio-gas storage pressure (see Fig. 3).

Figure 3: Simplified plant with floating and dome covers.

The necessity to store a larger quantities of bio-gas than that which can be stored using a normal gasometric cover can be satisfied by using external gasometers (spherical shaped and made of two or three adjustable volume membranes).

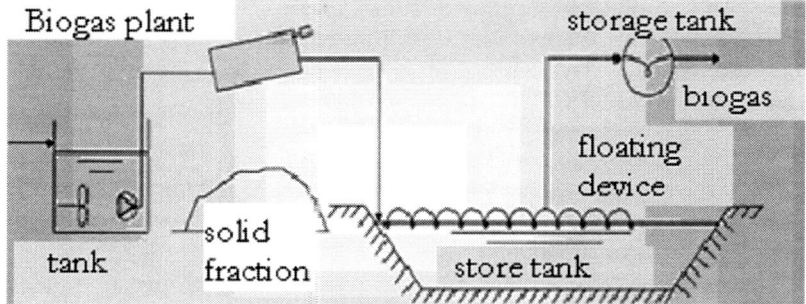

Figure 4: Scheme of a simplified bio-gas plant without heating.

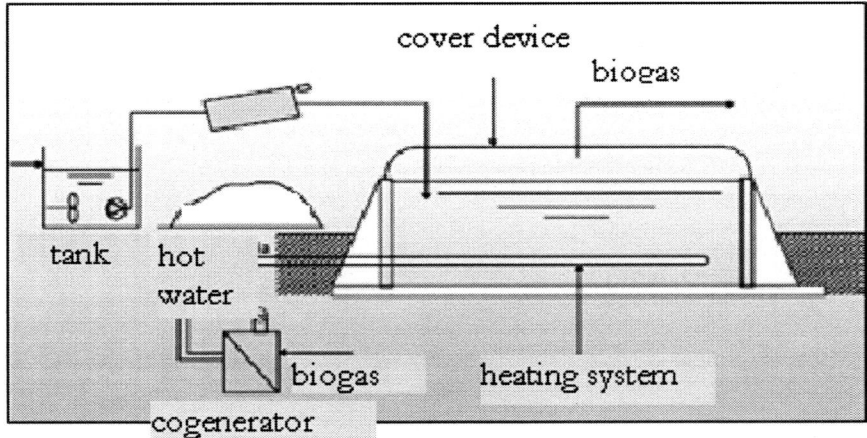

Figure 5: Scheme of a simplified bio-gas plant with heating.

Depending on a census taken at the end of 2004, more than 100 bio-gas plants were present in Italy; of these, 70 are simplified and low cost plants that have been realized by superimposing a plastic material cover over an effluent from intensive animal breeding in a storage basin [32].

We will now discuss two typologies of digestors which are more complex than the simplified plants discussed above: mixed reactors and 'plug flow' reactors.

2.1.1.2 Mixed reactors The mixed reactor is the more classic digestion typology. They are silo-shaped and are built in armoured concrete or steel. These reactors, working in the thermophile or mesophily regime, are equipped with a heating system that comprises a heat exchanger and they are insulated at the perimeter. Mechanical agitators at a low rotation regime allow mixing of the material to be digested. Depending on the number and the position of the agitators, the reactor can be completely or partly mixed. The gas produced by the anaerobic digestion process is retained by a gasometric dome placed on the top of the reactor and mainly made of a polymeric sheet that is protected by a steel cover or armoured concrete. This typology of the reactor allows treating liquid manures with a dry substance content that is lower than 10%, with medium permanence times, which are between 15 and 35 days depending on the composition of the substrate and the process temperature.

2.1.1.3 'Plug flow' reactors These reactors, equipped with a heating system, agitators and gasometers, allow the horizontal scroll of the liquid manure. They are only used on a small scale, because of technical and economic constraints that limit their volume to a maximum of 300–400 m^3. These systems, which are appropriate to treat liquid manures with a dry substance content of up to 13%, allow to obtain bio-gas yields that are higher than those obtained with mixed reactors, at equal temperatures.

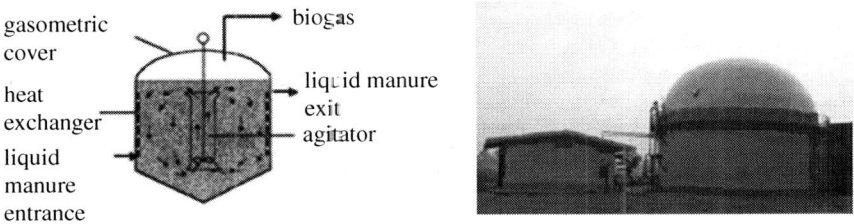

Figure 6: Complete mixing reactor.

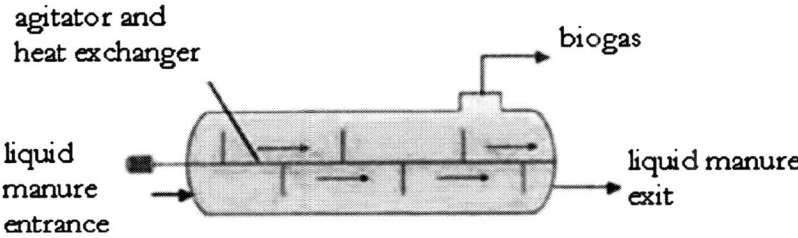

Figure 7: Schematic view of a 'plug flow' reactor.

Figure 8: 'Plug flow' reactors.

In addition to the two reactor versions analysed, which are more commonly applied, the market also offers other typologies of digestors that are more sophisticated and use high technology, which are particularly suitable for industrial effluent treatment at high organic load: reactors for contact, anaerobic filters and upflow anaerobic sludge blanket (UASB) reactors.

2.1.2 Co-digestion

The anaerobic digestion process can be realized simultaneously using more typologies of substrates (co-substrates): in this case, we talk about co-digestion.

The applicable co-substrates can be effluents from intensive animal breeding, agricultural residuals, bio-solids, agro-alimentary industry rejections, etc.

Co-digestion involves the use of complete mixing reactors, where the substrate is usually diluted to obtain a dry substance concentration that is between 8% and 15%. The use of co-digestion cannot be justified for higher bio-gas yields, but it results in the possibility of obtaining an extra income from the plant's administration [2, 37, 38].

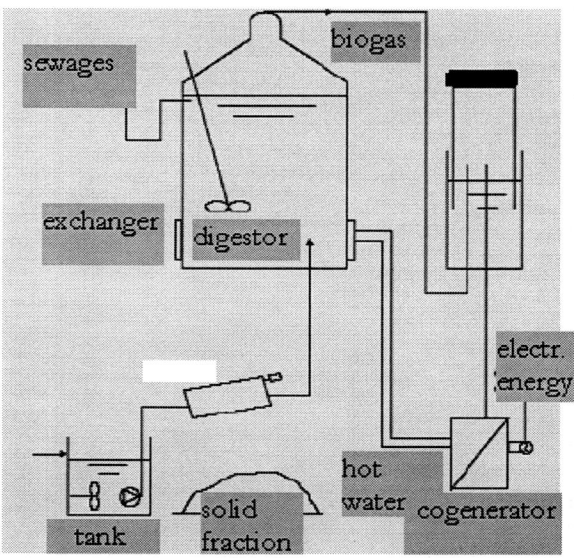

Figure 9:   Co-digestion plant scheme.

2.1.3 Bio-gas in solid rejections dumps

The management of rubbish and particularly the final rubbish disposal in controlled dumps with a high level of compaction of the material allow the establishment of anaerobiosis conditions that are necessary for bio-gas production after the decomposition of the organic content which is present in the rubbish. The collection of the gas produced, which happens through appropriate cargo systems, is necessary not only for the exploitation of energy but also for safety reasons (methane, in addition to being harmful to humans and vegetation by its greenhouse gas effect, is explosive in confined environments).

The collector system comprises a series of vertical shafts from which horizontal cracked piping spiders originate. The collection and removal of the gas are controlled by the pressure to which it is subjected inside the body of the dump.

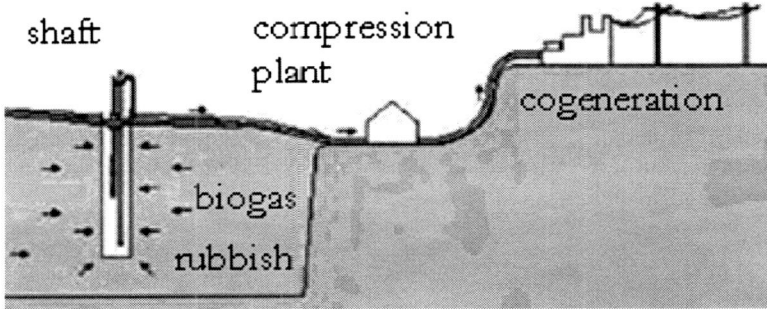

Figure 10: Production spinneret for energy produced from bio-gas in a controlled dump of solid urban waste.

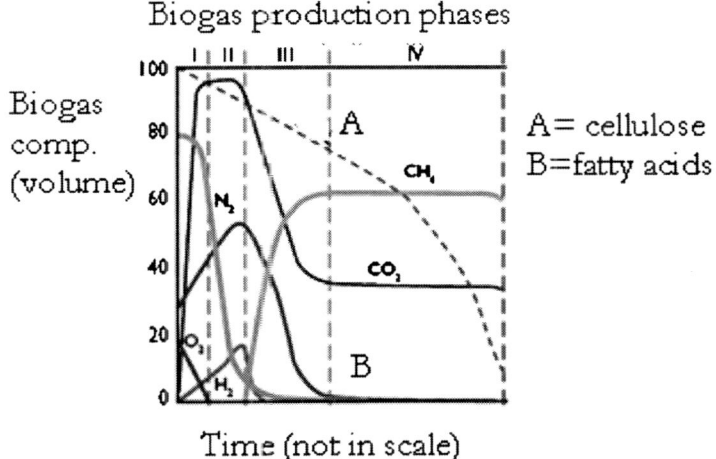

Phases	Conditions	Duration
I	aerobiosis	hours-1 week
II	anoxia	1-6 months
III	anaerobiosis mathenogenesis (instable)	3 months – 3 years
IV	anaerobiosis methanogenesis (stable)	8-10 years

Figure 11: Bio-gas production process phases and its composition.

2.2 Aerobic digestion

Aerobic digestion involves the metabolizing of the organic substances by micro-organisms, whose development is controlled by the presence of oxygen. These bacteria convert the complex substances into other simpler substances, freeing CO_2 and H_2O and result in high substrate heating (the layer under the biomass that is not yet digested by bacteria), which is proportional to their metabolic activity.

The heat produced can be transferred outside through a fluid exchanger. It is a process that is particularly suitable for liquid manures and industrial wastewater purification plants. In Europe, it is used (especially in Germany where strict rules were imposed for the destruction of pathogenic substances) in the thermophilic aerobic self-heated digestion process (auto-heated thermophilic aerobic digestion) for the treatment of the wastewater; more recently, this technology has also been used in Canada and the United States [3, 4, 35].

2.3 Alcohol fermentation

Alcohol fermentation is a biotechnological process which allows the production of bio-ethanol from simple carbohydrates (glucose, sucrose, mannose) and from long chain polysaccharides (starch, cellulose, hemicelluloses). Therefore, the raw materials necessary for obtaining this biocombustible can be derived from alcohol producing dedicated cultures both sacchariferous (sugar beet, sugar sorghum, etc.) and starchy (soft wheat and corn) as well as from lignocellulose residuals. The production spinneret for bio-ethanol is articulated in three sections: sacchariferous, starchy and lignocellulosic.

2.3.1 The sacchariferous section of the bio-ethanol production spinneret

The sacchariferous section is aimed at the energy conversion of the sugars that are obtained from sugarcane, sugar sorghum and sugar beet. From a technological point of view, the energy spinneret is articulated in the following phases: extraction of the sugars from the vegetable textiles, their fermentation and ethanol distillation. Generally, the fermentation is induced by the yeast *Saccharomyces cerevisiae*. It is used in bioreactors that reproduce the ideal conditions which favour its anaerobic metabolism (i.e. in presence of low oxygen concentrations), at a temperature between 5°C and 25°C and with a variable pH between 4.8 and 5.

Figure 12: Sugar beet and sorghum.

The separation of ethanol from the mixture that is obtained at the end of the fermentation phase (whose duration is around 72 hours) is achieved by distillation, exploiting the components present in the mixture. From the distillation, ethanol is obtained in concentrations equal to 95% of weight and with a residual water content of 5% [2, 30].

To obtain ethanol concentrations close to 100%, it is possible to resort, but with an increase of the process costs, to fractional distillation (concentration of ethanol equal to 99% of weight) or to the separation by pervapouration (concentration of ethanol equal to the 97% of weight). Fractional distillation involves the addition of benzene in the starting mixture.

The separation by pervapouration involves transporting the mixture that is obtained downstream of the fermentation in the vapour phase to filter it with appropriate selective hydrophile membranes.

2.3.2 The starchy section of the bio-ethanol production spinneret

In this section, bio-ethanol is obtained from raw materials that are rich in starch (soft wheat and corn). In this case, it is necessary to treat the starch by a hydrolysis reaction to make the glucose contained in it fermentable.

Wheat and maize grains are crushed and dehydrated to obtain starch paste. Starch is then gelated directly at a temperature of 175°C and a pressure of 2 atm. The hydrolysis is generally carried out using an enzyme called amylase, which has the ability to free the glucose molecules that are present in the starch chains. The temperature at which hydrolysis is carried out should be maintained below 60°C

Figure 13:   Wheat.

Figure 14:   Maize.

for a fermentable sugar yield of 80%. The remaining part of the starchy section is similar to the sacchariferous section [2, 30].

2.3.3   The cellulosic section of the bio-ethanol production spinneret

In this section, ethanol is obtained from raw cellulosic materials or from materials with high content of cellulose and hemicelluloses. Even though it is not possible to register the industrial scale production of cellulosic origin at the worldwide level, the possibility of using lignocellulosic residuals in this manner has initiated many research and development activities, particularly in the United States.

The main components of the lignocellulosic biomass are cellulose and hemicelluloses, and being made of fermentable sugars, they can be used to obtain ethanol. One of the critical points that characterize this spinneret section is the physical separation of cellulose and hemicelluloses from lignin. This can be achieved by chemical–physical (the most well known of which is called steam explosion, which applies saturated vapours at high pressure), chemical (with acids) and mechanical (with press systems) treatments. Subsequently, the cellulose and hemicelluloses are subjected to hydrolysis, which can be carried out in two ways: chemical or enzymatic.

The acid chemical hydrolysis can take place in only one step or in two different steps. In the first case, hydrolysis is carried out using concentrated (at a 77% concentration) sulphuric acid (H_2SO_4) which is added to the cellulose material in a ratio of 1.25:1 and at a temperature of 50°C. In the second case, dilute sulphuric acid is used: first, the hemicelluloses are attacked by H_2SO_4 at a concentration of 0.4% and at a temperature of 215°C. In enzymatic hydrolysis, the cleavage of the cellulose and hemicelluloses chains is achieved using enzymes that are called cellulases, which have been discovered in the micro-organism *Trichoderma reesei* and have also been subsequently identified in many other microbic groups.

Enzymatic hydrolysis is preferred to chemical hydrolysis. The hydrolysis of cellulose yields glucose molecules, which is an easily fermentable six-carbon atom sugar; the hydrolysis of hemicelluloses gives five-carbon atom sugars that are ethanol fermented with more difficulty. The total yield of bio-ethanol in the cellulosic section is still a matter of high concern, especially with regard to hemicelluloses.

2.4 Oil extraction and bio-diesel production

Bio-diesel can be produced by using as raw materials both oils extracted from oil cultures (vegetable oil extraction process: soy, sunflower, rape, etc.) and oils that are recovered (regeneration process of vegetable oils) from alimentary uses through separate collection systems. The products that are obtained upstream of the extraction and the regeneration can be directly applied as a combustible or directed to the transesterification process to obtain bio-diesel [1–3, 30].

Figure 15: Small scale biodiesel screw oil press.

2.4.1 Vegetable oil extraction

The steps involved in the extraction of vegetable oils from oil cultures are:

1. cleaning (with electrovalent elements or magnets);
2. grinding;
3. heating and conditioning (80–90°C, humidity 7%);
4. mechanical extraction (hydraulic press or strew-shaped) or chemical (solvents);
5. purification for depuration or refining (neutralization of the free fatty acids) if the oils are to be used for bio-diesel production.

The seeds derived from oil cultures are first cleaned, using magnets or electrovalent elements to remove additional materials or collect gross residuals, and are then decorticated. In the following grinding step, there is an outburst of the oils from the cells. Heating and conditioning, in the temperature range 80–90°C and at a humidity of 7–10%, promote the lysis of the cells, the diffusion of the seeds' fat material and the separation of the proteic components. Oil extraction from milled seeds can be done by mechanical or chemical techniques. Mechanical extraction uses a screw or hydraulic press, and it leaves an unextracted residual fat content equal to 5–12%. Chemical extraction, characterized by an unextracted residual fat content equal to 1% (with a seed–solvent ratio equal to 1:18, reaction environment temperature of 50°C and contact times equal to 2 hours for rape seeds and 1 hour for sunflower seeds), involves the use of organic solvents (such as trichloroethylene, hexane, carbon sulphur). Chemical extraction can be done in a discontinuous manner (batch), which is the preferred option for plants that treat at least 250–500 t/day. Chemical and thermochemical extraction can be integrated with each other. Despite a high investment cost, the yield is close to 100%. In this technique, the milled material is first subjected to mechanical extraction, which leaves a residual fat content of 20–24%, and, subsequently, chemical extraction is carried out. The yield of raw oil obtained from the extraction process is variable, from rape and sunflower seeds 36–38% of oil weight is extracted. If we want to convert the raw oil obtained into bio-diesel, it must be subjected to a purification step that can be performed depending on two modalities: purification and refining. The two processes, which are preceded by a centrifugation step, are diverse in terms of the qualitative levels, which is higher in the refining step. Depuration is directed to the removal of impurities (waxes, resins, pigments and mucilages) present in the raw oil and it is carried out using sulphuric acid, salt water solutions or by percolation using absorbing grounds. Refining removes impurities working in salt solutions with sulphuric acid or citric acid. Furthermore, refining reduces the acidity of the raw oils for the neutralization, which can be done in a physical (at 240–260°C and in conditions of vacuum at −1 mbar) or chemical manner (working with sodium hydroxide at a temperature of 60–80°C and at atmospheric pressure). The best quality of the refined oil, compared to the depured oil, is the reduction of the acidity of the raw oils. At the end of the purification step, the vegetable oil yield is about 34.4%. The main by-product of the vegetable oil extraction is the proteic panel, which is used in zootechny as animal feeds [2, 14, 24].

2.4.2 Vegetable oil regeneration

The alimentary origin-exhausted vegetable oils that can be used for energy valorization are those that are obtained from industrial processes (from ovens and fryers) and from domestic users (frying oils and oil for food conservation). Before being transformed into bio-diesel, the oils of alimentary origin must be subjected to the regeneration process which involves the following steps: removal of the rough impurities through subsequent filtrations, neutralization and dehydration. The exhausted regenerated vegetable oils can be used in a similar manner as the oils obtained from dedicated cultures and from this point they follow the same spinneret [2].

2.4.3 Transesterification

The reaction that is used for the synthesis of bio-diesel is known as transesterification, a process in which oils react with methanol, in the presence of a catalyst, to form methyl ester (bio-diesel) and raw glycerine, as a secondary product. In other words, the reaction induces the breakdown of the triglycerides molecules that make up the vegetable fats to obtain the methyl ester of the fatty acids mixture. The main result of this process is the reduction in the viscosity of the starting oils, making them compatible with certain energy uses and particularly for use of the bio-diesel as a auto traction fuel.

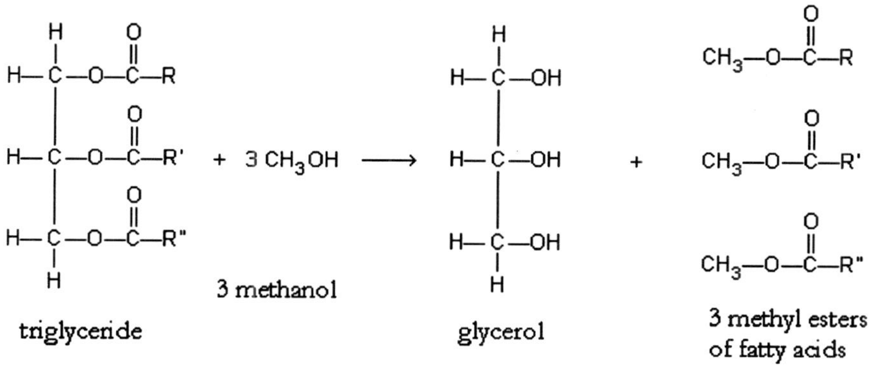

$$\begin{array}{c}
\text{H} \quad \text{O} \\
| \quad \quad \| \\
\text{H--C--O--C--R} \\
| \\
\text{H--C--O--C--R'} + 3\,CH_3OH \longrightarrow \\
| \\
\text{H--C--O--C--R''} \\
| \\
\text{H}
\end{array}
\quad
\begin{array}{c}
\text{H} \\
| \\
\text{H--C--OH} \\
| \\
\text{H--C--OH} + \\
| \\
\text{H--C--OH} \\
| \\
\text{H}
\end{array}
\quad
\begin{array}{c}
CH_3\text{--O--C--R} \\
CH_3\text{--O--C--R'} \\
CH_3\text{--O--C--R''}
\end{array}$$

triglyceride   3 methanol   glycerol   3 methyl esters of fatty acids

Figure 16: Bio-diesel production reaction.

The simplified representation of the entire process is as follows:

1000 kg of refined oil + 100 kg of methanol = 1000 kg of bio-diesel
+ 100 kg of glycerine

There are several plant solutions for the realization of bio-diesel from vegetable oils, where the medium yield of conversion is equal to 98%. The factors that affect the choice of the technology to be adopted are the quantity to be treated, the periodicity with which the raw materials are available and the quantity of oils at the plant entrance [2, 3].

There are three main technologies that are distinguished in terms of the process temperature and pressure [2, 14]:

- *Environment temperature plant:* The process takes place at environmental temperature and atmospheric pressure, using sodium or potassium hydroxide as the catalyst. It is appropriate for batch treatment (discontinuous), with a bio-diesel production capacity of up to 3,000 t/year. The time of reaction is 8 hours.
- *Medium–high temperature plant:* The process takes place at atmospheric pressure and at a temperature of 70°C, using sodium or potassium hydroxide as the

catalyst. It is appropriate for continuous or batch treatment, with a production capacity of up to 25,000 t/year. The reaction time is 1 hour.
- *High temperature and pressure plant:* The transesterification takes place at a pressure of 50 MPa and a temperature of 200°C. This technology, involving higher installation and management costs, is justified for production capacity higher than 25,000 t/year, both in continuous and in batch treatments. This technique, compared to the other two techniques, allows the treatment of high acidity oils (up to 4%) because it uses phosphoric acid for acid catalysis.

All the technological solutions described above involve the recovery of excess methanol for the vacuum distillation (stripping) and its reuse at the start of the plant.

The main by-product of the transesterification process is a glycerol that has a high economic value in pharmaceutical and cosmetic applications.

3 Thermochemical conversion

The thermochemical conversion processes are [3, 36]:

- direct combustion,
- gasification and
- pyrolysis.

3.1 Direct combustion

Among the several processes for the thermochemical conversion of biomasses, direct combustion is, without doubt, the most ancient and mature technology. Despite this, many research studies are being continuously carried out with the aim of developing this technique further, making it more and more efficient and with lower environmental impact. The combustion process allows the transformation of the chemical energy in the biomass into thermal energy, through a series of chemical–physical reactions. When a biomass enters the combustion room, first, it is subjected to drying; subsequently, as the temperature increases, pyrolysis, gasification and combustion processes occur. With appropriate combustibles/air profiles, the biomass decomposes and volatilizes, freeing a carbon residual (cinders), which is mainly made of mineral inert compounds. The end result of these processes is the production of heat which is recovered through heat exchangers in which the thermochemical energy is transferred to other vector fluids such as air or water. The quantity of thermal energy that is contained in the biomass is a function of type of the cinders and the humidity content and it is generally defined by the lower calorific power [4, 28].

The different combustion technologies used are [4, 28, 30]:

- Grate shaped (fixed or moving), fundamental element in addition to the thermal reaction, also for the removal of the cinders; fixed systems are generally used for combustors of small size. For industrialized plants, moving grates are used as they facilitate the handling, the mixing of the combustible and the removal of its cinders; such grates can be of different types: horizontally or vertically vibrating, belt, rotating, steps, rolls, etc., and, in some cases, they are cooled with air or water to allow a higher specific heat load.

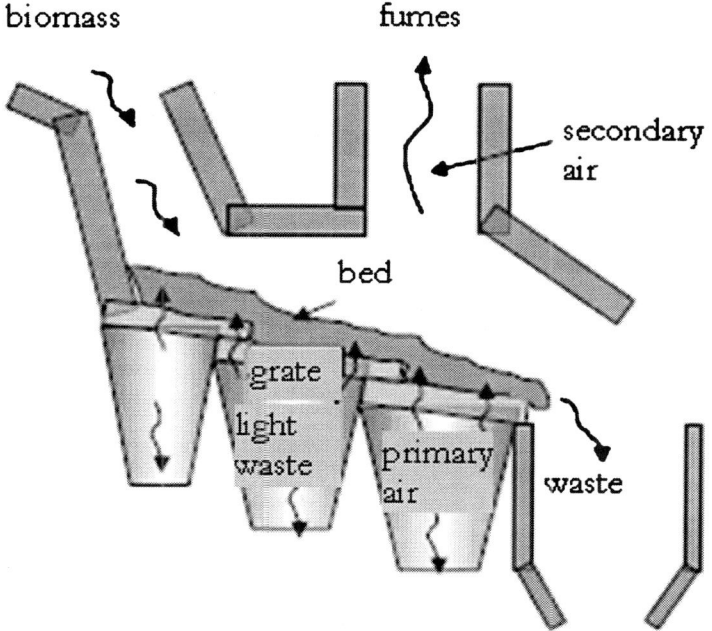

Figure 17:   Grate oven working scheme.

Figure 18:   Switched off oven moving grate.

- In suspension, appropriate for powdery and light biomasses, such as rice husk, sawdust, wood dust and chaff, in which the biomass is fed in the upper part of the combustor where it burns and falls on the grate beneath, whose main function is to remove the cinders.
- Rotating drum, used for applications in which the combustible has thermophysical characteristics, particularly, poor and high polluting load characteristics. During combustion, the biomass is continuously remixed by the low drum rotation and the direction of the combustion product can be either in the same direction or in the opposite direction to the biomass progress direction.

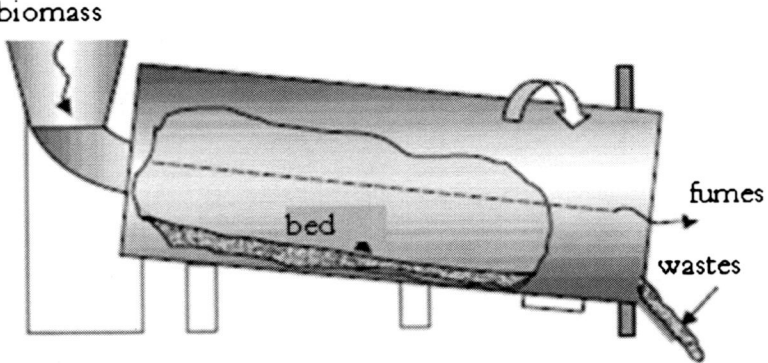

Figure 19:   Rotating drum oven working scheme.

- Double stadium shaped, in which, first, gasification and material pyrolysis take place in one room and complete combustion of the gasified products takes place in another room, resulting in the transfer of a major portion of the energy to the operating fluid.
- Fluid bed, in which several kinds of biomass can be treated, including selected urban solid rejections even with a high humidity percentage (>40%). The combustion room is partially filled with inert material such as sand or alumina, which is fluidized from the primary combustion air to establish a 'boiling bed' or if there is higher air speed and material dragging, the so-called 'recirculator bed', which is recovered and re-fed into the combustion room. In addition to the inert material, even the material that allows to change the environment conditions in which the combustion takes place can be fed into the combustion room: in fact, if polluting combustibles with acidic compounds or containing low-flux cinders are present, limestone or dolomite can be used to neutralize the polluting acids and to avoid fusion of the cinders in the combustor's operating conditions.

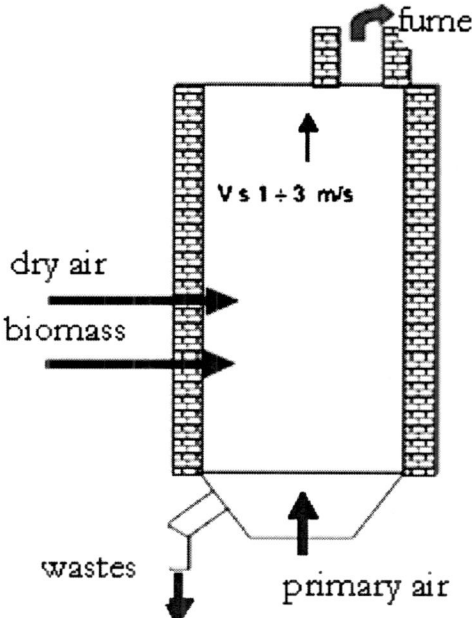

Figure 20: Fluid boiling bed.

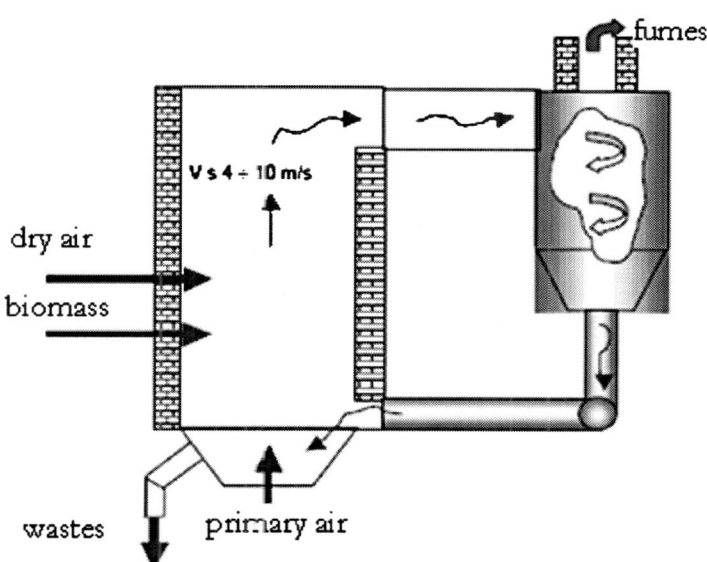

Figure 21: Fluid recirculated bed.

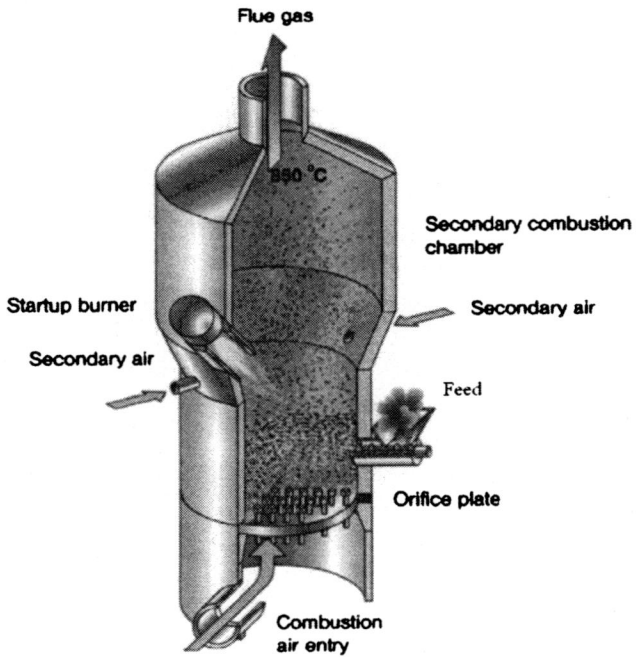

Figure 22: Fluid boiling bed.

The use of the combustion devices facilitate the recovery of the maximum amount of the energy developed during the process. This recovery can take place in a direct manner through the device's walls (stoves) or in an indirect manner through a vector fluid (boilers). The presence of the heat recovery sections is not only convenient from the energy and economic point of view, but it is also necessary to reduce the temperature of the fumes that are emitted from the combustion room (temperatures of 1200°C can be reached) as it is possible to bring down their temperature (to not higher than 300°C). The combustion devices show different constructive characteristics depending on their usage, whether it is meant for the civil, agricultural or industrial sector. The devices that are used in the civil sector (environment heating) include many models that are used commercially at both the national and the European level, and are classified as follows [28]:

- thermal wood kitchen, which are only used for monofamiliar purposes, both for heating environments and for cooking food, having a global yield of 70–75%;
- thermal wood chimneys, also meant for monofamiliar use, with water or air exchangers and which have a medium efficiency equal to 50%;
- wood little-medium power boilers (20–300 kW_{th}), having a medium variable efficiency between 60% and 80%, which allow the heating of single habitable units or small residential plants; they are equipped with smaller-sized fixed grates and involve manual loading of the combustible, whereas for the higher

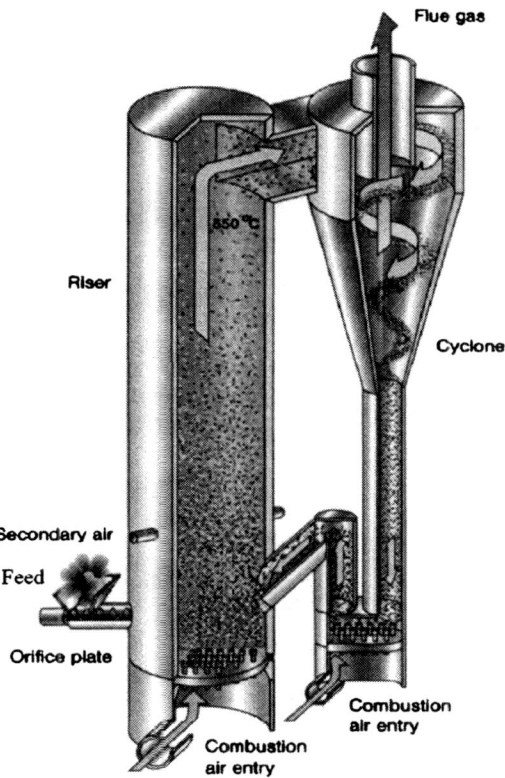

Figure 23: Fluid recirculated bed.

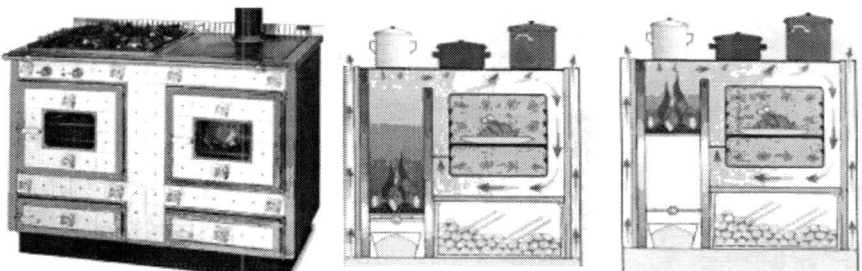

Figure 24: Wood thermal kitchen.

powers, there are loading hoppers, feed devices, fixed and moving grates, cinder and dust fellers before the fumes are discharged to the chimney evacuation systems. The boilers for water heating are of the fume pipe type, in which the hot combustion gas passes through the tube bundle which is immersed in water to which the heat is transferred.

Figure 25: Wood burning stoves.

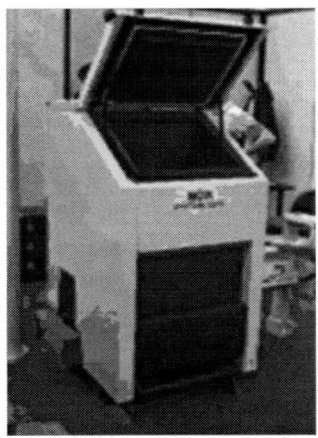

Figure 26: Little power for the combustion of wood log boiler.

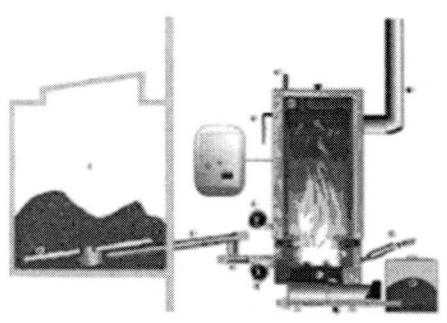

Figure 27: Little-medium power chips or pellet powered boiler.

For the agricultural sector, there are particularly interesting large room combustors and the moving grate combustors, which are equipped with straw bale feeding systems, tree pruning residuals, agro-industrial working residuals, etc. The combustors must be planned appropriately to ensure good working with biomasses, which are characterized by wide variations in their humidity levels. The most frequent applications, and in many cases the most economically viable, which are registered in this sector are the drying of agricultural products and the greenhouses and buildings for piggish and poultry cattle heating, in addition to the normal domestic heating.

The power of thermal devices is generally between 200 and 2,000 kW_{th}. Even in this case, the heat exchanger is the fume pipe type which has described previously [28].

In the industrial sector, there are many biomass direct combustion agro-forest or agro-industrial application for the urban solid rejections (RSU) and the industrial wastes. These applications allow the production of heat that is used in the production cycle for the generation of electrical energy and cogeneration products (simultaneous production of electrical and thermal energy). These plants comprise the following sections:

- biomasses stocking, which can have dimensions that can guarantee the supply of combustible for some days or very long periods (also for some months), if biomasses of a seasonal nature are processed;
- additional pre-treatment, which consists of reducing the biomass sizes and humidity to the specific requirements of the combustion system;
- feed line which is equipped with appropriate flow controls;
- combustor with the characteristics described previously;
- energy recovery, through fume pipe systems if the vector fluid is low pressure air or hot water, water pipes are used if it is necessary to have water at overheated pressure or vapour, diathermal oil, and additionally the exchanger (not in high power plants).

If meant for electrical energy production plants, it is necessary to introduce additional components such as the vapour turbine and linked to it the electro-generator, the vapour condensator, the degasser and several thermal recuperators for the optimization of the thermal cycle. To drive the vapour turbines, vapour should be generated at medium–high pressure. The power of the plants that produce only thermal energy can vary from some hundreds of kilowatts to some tens of MW_{th}: the limit of the larger sized biomass industrial plants with wood spinnerets or other types of biomasses with both technical and organizing managerial character. Even the number of yearly working hours is often a limiting factor for the economic investment gain, when compared with traditional feed combustible plants, because they generally show low investment costs against high energy costs.

The construction of plants for electrical energy production and for cogeneration by combustion is more economically advantageous only when the biomass is available in large quantities that are ideally located geographically placed and time distributed because the biomass has a low energy density, 10 times lower than that of petrol. This can be achieved only after a considerable reduction in the transport and stocking incidence of the quantities that are necessary for a central working, whose typical power is generally in the range 3–10 MW_e. To give an idea of the biomass requirements for this kind realization, the need for 1 kg of biomass to produce 1 kW h of electrical energy should be considered.

The main energy parameter that is applied to evaluate the plants is the global net yield, which is given by the percentage ratio of the energy available for external users and that introduced by the combustible in the energy production plant, which are expressed in the same unit measures, net necessary consumptions for the working of the same plant [28, 30].

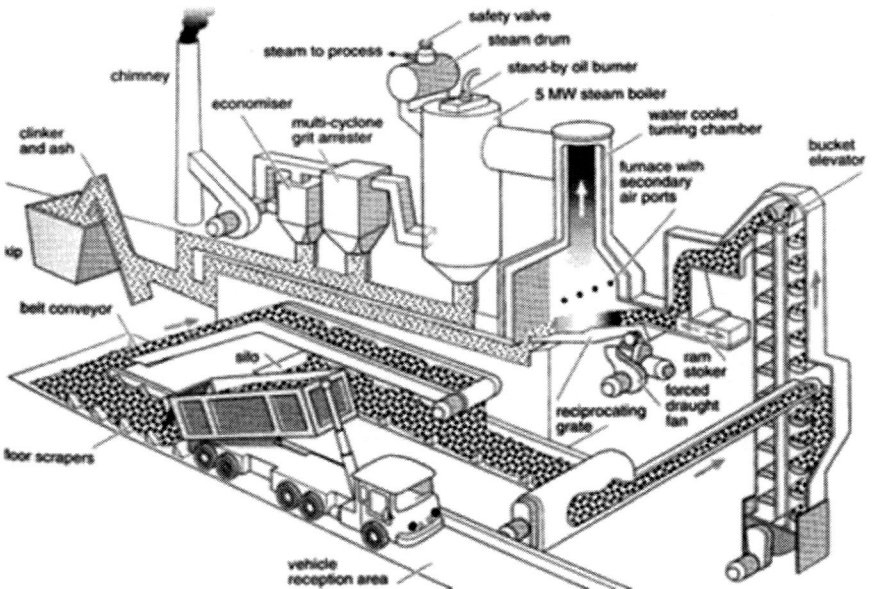

Figure 28: Biomass combustion plant.

3.2 Gasification

Gasification is a thermochemical conversion process that transforms a solid combustible into a gaseous combustible which shows ease of complete combustion without the need for excess air, ease of turning on and ease of transport and cleaning of the combustion. The disadvantage is the energy expense that is required for the gasification process. The gas obtained from the gasification of biomasses is called producer gas; it is composed of a mixture of carbon monoxide, hydrogen, carbon dioxide, methane, hydrocarbons (ethylene, ethane), vapour, nitrogen (in air gasification); it also contains several pollutants such as cinder and char particles (agglomerates of a complex nature which are mainly composed of carbon), tar (complex mixture of condensable hydrocarbons) and oils. Producer gas can be obtained through partial combustion of the biomass (using air or oxygen) or through pyrolytic gasification (using vapour). Gasification in air results in a low calorific power gas (5.5–7.5 MJ/N m^3), whereas by oxygen and vapour gasification a medium calorific power gas is obtained (11 MJ/N m^3 and 10 MJ/N m^3, respectively). For pyrolytic gasification (or indirect heating), an external heat supply is necessary.

All gasification processes involve, with different modalities depending on the technology applied, the following four steps: drying, pyrolysis, oxidation and reduction.

Drying involves the evaporation of the water content, introduced in the reactor, in the biomass. Pyrolysis is the decomposition of the biomass at high temperatures without exposure to oxygen: the products of pyrolysis are gas (containing tar in the vapour state, in addition to substances such as methane, hydrogen, carbon monoxide, carbon dioxide and hydrocarbons with a few carbon atoms) and char. In the oxidation phase, the exothermic reactions, which provide the heat required for the reduction reactions (endothermic), take place from which the constituents of the producer gas originate. There are multiple technologies by which it is possible to realize the gasification and which are mainly distinguished by the manner in which the biomass is brought into contact with the gasifying agent. We can distinguish two main reactor categories: fixed bed (updraft, downdraft, crossdraft) and fluid bed (boiling and circulating). Most of the gasificators in use are of the downdraft type [2, 46].

3.2.1 Fixed-bed gasificators

These gasificators represent the most tested gasification technology. They show the highest limited dimensions and low reaction speed, although their use is limited to the smaller powers. The feed material must have uniform granulometry and a low fine particle content, to avoid overloads and allow 'empty space' which is enough for the passage of gas through the bed.

3.2.1.1 The updraft or counter-current gasificators The reactor is made of steel cylinders coated inside with refractory material. In the upper part of the reactor, there is the biomass feed and the exit for the producer gas, whereas in the lower part there is a grid that functions as a support for the solid material bed. The grid allows the passage of both air that enters from the bottom and cinders that fall down and collect at the bottom. This gasificator typology has the following advantages:

- constructive and working simplicity;
- high combustion capacity of char, whose final residual is minimal;
- optimal thermal exchange between the opposite currents of biomass and the producer gas, which results in low exit temperature of the producer gas and therefore a high thermal efficiency;
- efficient drying of the combustible due to the internal thermal exchange; this allows the use of combustibles with high humidity levels (up to 60%).

The fundamental limitation of this technology is the high tar content in the producer gas. The tars mainly originate during the pyrolysis, and in this type of gasificator, the pyrolysis gas, containing tars, combines with the producer gas

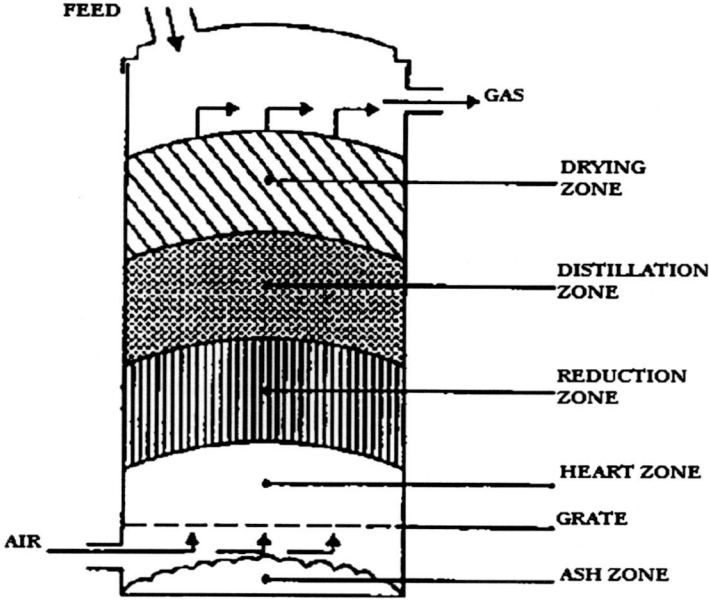

Figure 29: Updraft gasificators.

without being burnt before The tars can pose considerable problems in the producer gas feed plants; in fact, they condense easily and provoke overloads. This is very important if the gas is used in a boiler, whereas in case of use in turbines or engines, an accurate cleaning of the gas is necessary. These gasificators are characterized by a maximum load of 4 t/h of dry biomass [2, 46].

3.2.1.2 Downdraft or equi-current gasificators In the downdraft gasificators, the current of producer gas is descending and so it is concordant with the current of the solid combustible. The gas exits the reactor from the bottom. Generally, they have a V-shaped throat above which the oxidation zone is located. The purpose is to create a compact zone at high temperature where the pyrolysis gas is generated and to realize the cracking of the tars (decomposition into lighter products); the air is allowed to directly enter this area through a central feed pipe or through nozzles that are placed on the groove walls.

The main advantage of this type of gasificator is the low tar content in the producer gas. The limitations are:

- high content of solid particles in the producer gas, because the pyrolysis gas passes through the oxidation zone where it collects cinders and dust;
- the presence of grooves can pose overload problems;
- the humidity of the biomass must be lower than 35% because the internal drying is less efficient compared to the updraft gasificators;
- the relatively high temperature of the gas at the exit which reduces the thermal efficiency.

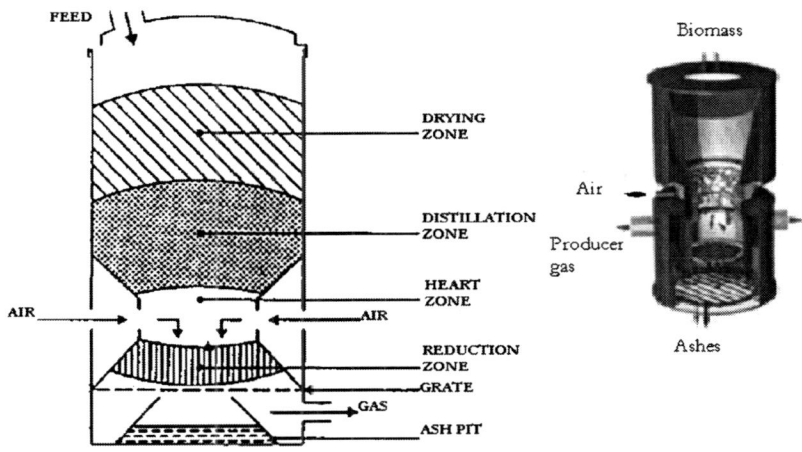

Figure 30: Downdraft gasificator.

This type of gasificator, characterized by a maximum load of 500 kg/h of dry biomass, is applied in small-scale applications up to 1.5 MW$_{th}$ [2, 46, 48].

3.2.1.3 Crossdraft gasificators The working of crossdraft gasificators is similar to the other two gasificators, but in this case the combustible is injected from above, the oxidant enters transversally and the producer exits laterally.

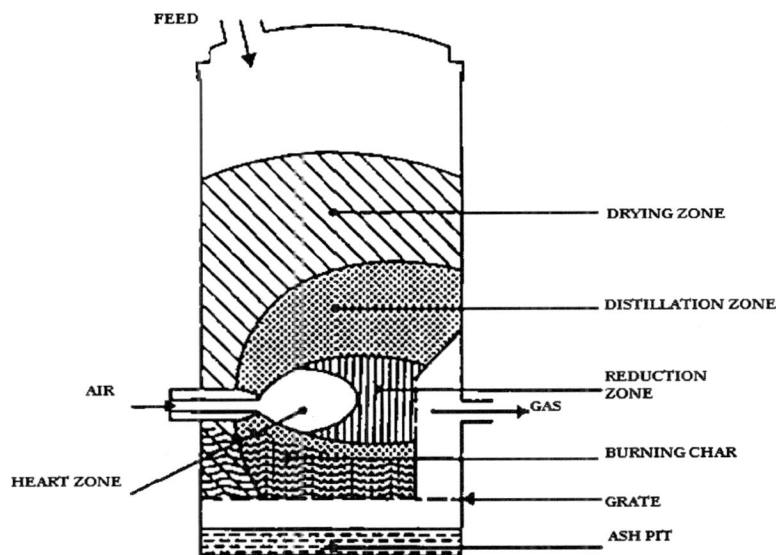

Figure 31: Crossdraft gasificator.

The disadvantage of crossdraft gasificators that results in it not being used is the reduced capacity of tar conversion [46, 48].

3.2.2 Fluid-bed gasificators

A fluid bed is a suspension of solid particles in an ascendant current of gas. The gas is introduced at pressure from the bottom of the reactor, whereas the particles enter from above. When the solid remains in suspension, we talk about the fluidization condition that is reached for a determinate speed of the gas in which the fluid bed, made of a solid phase and a gaseous phase, acts as a liquid. The application of the fluid-bed technology to gasification ensures very good mixing between the biomass (reduced into small particles) and the aerating agent, improving the reaction speed. We can also introduce inert fluidizing material (silica sand, alumina, refractory oxides) in the bed with the aim of equalizing the temperature so as to facilitate the heat transfer between the particles and the combustible.

The fluid-bed reactors, in contrast to fixed-bed applications, are characterized by a uniform temperature in the reactor (typically 800–850°C). Using the fluid-bed technology, we obtain a tar content in the producer gas that is intermediate between that obtained from the updraft gasificator and the downdraft gasificator.

The gas exiting from the reactor shows a higher content of solid particles (char, cinder, sand) [2].

3.2.2.1 Fluid boiling bed gasificators (BFB, bubbling fluidized bed)
In the fluid boiling bed gasificators, the height of the bed is limited (1–2 m) and the speed of the gas is 0.8–2 m/s (minimum speed necessary to keep the solid phase in suspension). Over the bed, there is a region where only the gaseous phase is present. Inside the bed the gas bubbles are formed whose movement on the surface resembles the phenomenon of a liquid that is boiling; this provokes an internal agitation that leads to further mixing of the phases. These reactors show higher temperatures compared with the temperatures observed in the fluid circulated bed reactors, which results in a lower tar content in the producer gas but also presents a bigger danger of cinder fusion.

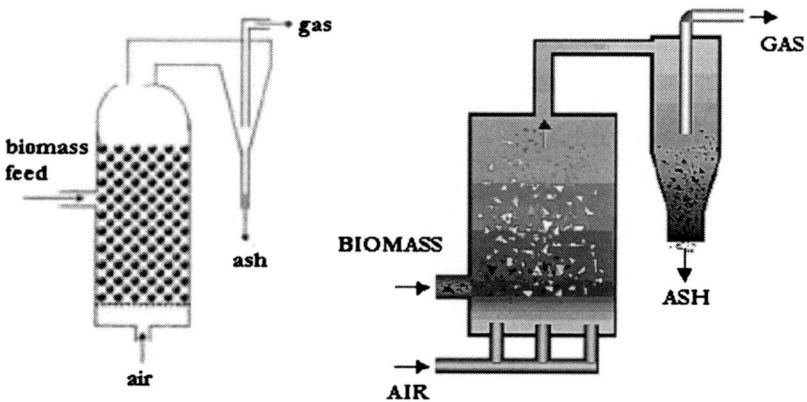

Figure 32: Fluid boiling bed reactor.

The main characteristics of this type of reactor are:

- high reaction speed;
- better temperature control than the fixed-bed reactors;
- high solid particle content in the producer gas;
- low tar content in the producer gas;
- maximum loads of the order 10–15 t/h of dry biomass;
- flexibility in terms of the granulometry of the biomass feed;
- for equal dimensions, the fluid-bed gasificators have higher powers than the fixed-bed gasificators;
- ease of turning on and switching off;
- catalysts can be added to the bed for the cracking of the tars;
- loss of carbon in the cinder.

The fluid boiling bed atmospheric gasificators are appropriate for different types of biomass and for applications with medium and small powers up to 25 MW_{th} [2, 46, 48].

3.2.2.2 Fluid circulating bed reactors (CFB, circulating fluidized bed) The reactor has heights reaching 8 m and it has a limited diameter. Given that the gas speed is high (>4 m/s), the solid particles (char and sand) are dragged until they go out of the main column, to be, then, separated from the gas through a cyclone and introduced again at the bottom of the reactor.

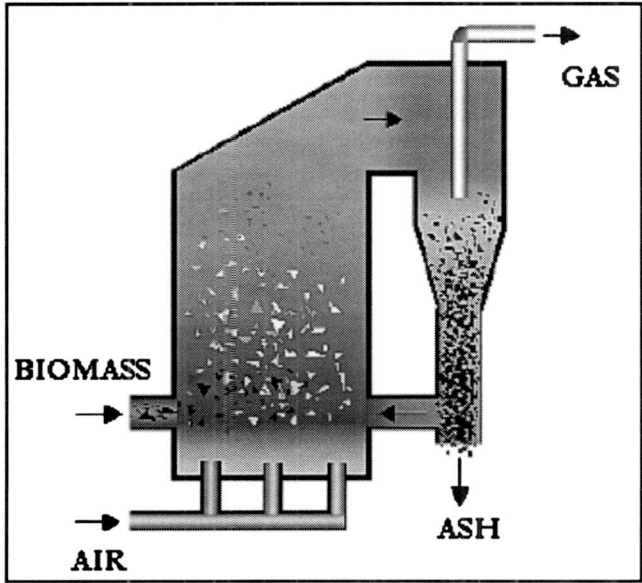

Figure 33: Circulating fluidized bed gasificator.

Starting from the bottom, the bed shows three different areas:

- dense phase: it is characterized by a high density and by the formation of gas bubbles;
- intermediate phase: unstable area with regions of different densities;
- dilute phase: the mixing of the solid in the gas is homogeneous and the density is low.

The main characteristics of the circulating fluidized bed gasificators that differentiate them from the boiling fluid bed gasificators are:

- for low powers, they involve higher costs compared with the boiling fluid bed gasificators;
- difficulty in the realization of cracking of the tars inside the bed;
- utilization for biomass loads that are higher than 15 t/h.

The circulating fluid bed atmospheric gasificators are appropriate for a huge variety of biomasses, with powers that vary from few MW_{th} up to 100 MW_{th}. This technology is more appropriate for large-scale applications [2, 46, 48].

3.2.2.3 Dual bend gasificators for pyrolytic gasification In this case, the gasification does not take place through partial oxidation but through indirect heating of the biomass (pyrolytic gasification). The plant is composed of two fluid-bed reactors: a circulating fluidized bed gasificator and a combustor (boiling or recirculated fluidized bed). In the gasificator, the heat required for the decomposition is obtained from the freewheeling sand in the plant which is heated in the combustor. Vapour is used as the fluidized gas. The producer gas that exits from the gasificator drags the sand particles and char that are separated by a cyclone and carried to the combustor, where the char is burned. The heat generated is absorbed from the sand that is dragged out of the combustor by the waste gas. A second cyclone provides for the exhausted gas sand, allowing its reintroduction into the gasificator where it transfers the absorbed heat to the biomass.

The process is particularly complex and the huge dimensions makes the realization of the plant particularly difficult, because of the high investment costs required. The main advantage of this technology the use of vapour which allows producing a medium calorific power gas without the use of oxygen. Given that a part of the char must be used for the combustion, there is low carbon conversion in the gas. In fact, this technology shows a high tar content in the producer gas [2]

3.2.2.4 Pressurized fluid bed gasificators When the producer gas is applied as a combustible in gas turbine plants, it should enter the combustor at high pressures (10–25 bar). If the gasification takes place in an atmospheric reactor, the gas must be cooled and compressed, spending energy. One solution to this problem is represented by the pressurized fluid bed gasificator which allows obtaining high pressure gas directly.

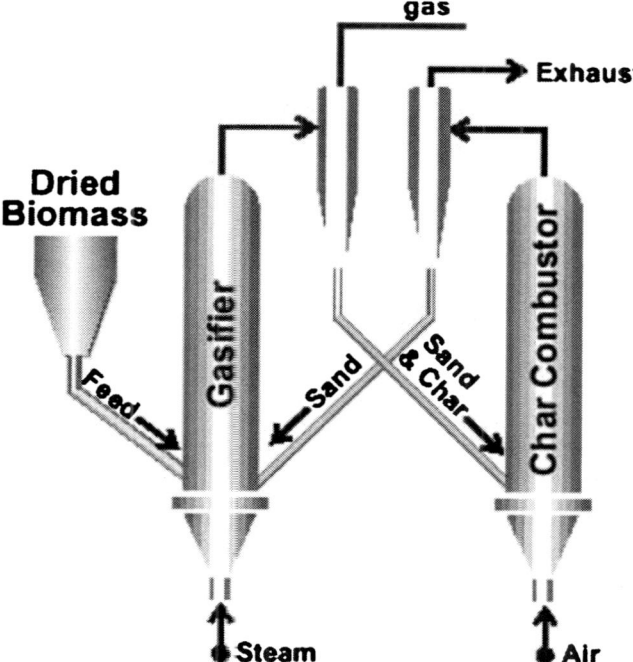

Figure 34: Dual bed gasificator.

The use of the pressurized fluid bed gasificators has the following advantages [2]:

- low internal energy consumption (because the gas notneed not be compressed);
- at high pressures, the tendency of the cinder to sinter is reduced;
- more compact dimensions compared with the atmospheric gasificators;
- low danger of condensation because the gas case is not cooled before use.

The disadvantages are:

- difficult to feed the biomass into the reactor;
- high investment costs;
- hot cleaning gas devices that are expensive and still in the development phase.

3.2.3 Producer gas applications

The production of electrical energy represents the most interesting gas use modality. Currently, the vapour plants in which the direct combustion of the biomass takes place are the most widely used technology to produce electrical energy from biomass.

A possible use of the producer gas is the co-firing (co-combustion) with traditional combustibles in vapour plants. In this way, there is a considerable saving of fossil fuel. Furthermore, the investment costs are low; resorting to co-firing requires only little modifications to the existing plants.

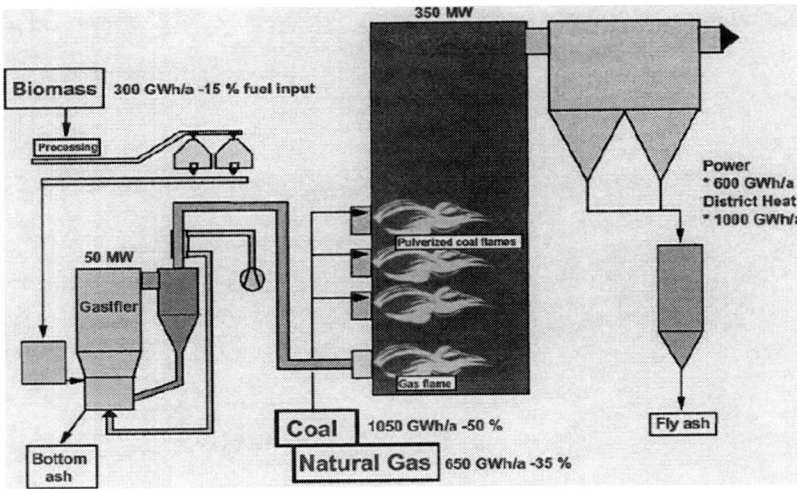

Figure 35: Example of co-firing (coal, natural gas and producer gas): scheme of Rankine cycle vapour generation plant for the production of electrical energy (200 MW$_e$) and heat (250 MW$_{th}$) located at Lathi (Finland).

The most promising technology is represented by the combined gas–vapour cycles that are integrated with gasification (IGCC, integrated gasification combined cycle). In systems with powers of tens of megawatts, global earnings as high as up to 50% have been achieved. The plans proposed until now are essentially of three types:

- fluid bed atmospheric gasification with air and cleaning of the gas by humid cleaning;
- fluid bed pressurized gasification with air and hot cleaning of the gas;
- indirect heating atmospheric gasification with humid cleaning.

Commercially, the IGCC systems are not yet competitive, but they can become competitive in the short term, especially due to the possibility of realizing co-generation plants [2, 46, 48].

3.3 Pyrolysis

The process of pyrolysis consists of a thermochemical conversion that allows transforming the organic substance into final fuel products (solid, liquid, gaseous). Pyrolysis takes place in the absence of oxidizing agents, or with a limited presence of these agents so that the oxidation reactions can be neglected. The heat required for the evolution of the process can be indirectly supplied through the reactors walls (transport of heat for convention and irradiation) or directly by recirculating a heating tool in the bed (heat transport for conduction) [44].

Figure 36: IGCC plant in Varnamo in Sweden.

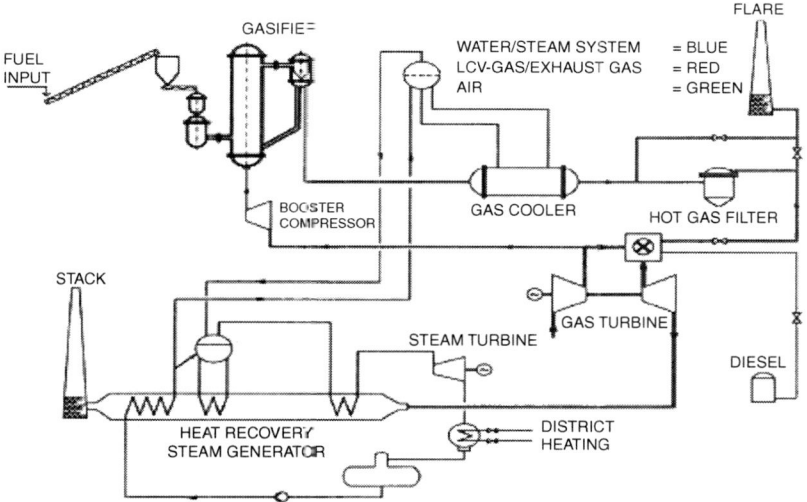

Figure 37: IGCC Varnano plant scheme.

The products of pyrolysis, although they differ depending on the feed material, can include the following [44]:

- a fuel gas having a medium calorific power (13–21 MJ/N m^3), mainly made of CO, CO_2 (if oxygen is present in the basic material), H_2 and light hydrocarbons (both saturated and unsaturated);

- a liquid product (obtainable from the condensation of the vapour phase) that is separated into two phases: an aqueous phase containing low molecular weight organic species that are soluble and a non-aqueous phase that is mainly made of organic molecules and oils with high molecular weight, called tar or bio-oil;
- a solid carbon product (char) and the cinders.

The most common modalities for execution of the pyrolysis process are [30, 44]:

- 'Carbonization', the most ancient and well-known pyrolysis process, which takes place at temperatures between 300°C and 500°C. From this process, only the solid fraction (vegetable carbon) is obtained and therefore the other fractions can be minimized.
- 'Conventional pyrolysis', which takes place at temperatures lower than 600°C, with moderate reaction times. From this process three fractions in about the same proportions are obtained.
- 'Fast pyrolysis', which takes place at temperatures between 500°C and 650°C, with brief reaction times. This process favours the production of a liquid fraction up to 70–80 % of the feed biomass weight.
- 'Flash pyrolysis', similar to fast pyrolysis but which take place at temperatures higher than 700°C and have reaction times that are lower than the former. This process allows the production of a liquid fraction up to 80% of the feed biomass weight, but with a composition variation that is more restricted than that of the fraction obtained by fast pyrolysis.

Table 1: Brief description of the pyrolysis processes [4].

Typology	Temperature (°C)	Characteristics
Carbonization	300–500	Only recovers solid fraction (coal)
Conventional pyrolysis	< 600	Three fractions of same proportion
Fast pyrolysis	500–650	Production at 70–80% of liquid fraction
Flash pyrolysis	>700	Production at 80% of liquid fraction

The main parameters that influence the process are [44]:

- temperature and pressure;
- speed of feed heating;
- dimensions and shape of the biomass to be treated;
- presence of additional catalysts;
- residence times of the solid phase and volatile phase in the reactor.

The products of pyrolysis can be used for the following purposes [44]:

- The gas: It can be burnt to give heat to the reactor involved in the pyrolysis or it can be applied as a fuel in turbo-gas or internal fuel engines.

- The tar: In most of cases, it is not directly applicable as a fuel because of its high viscosity and acidity due to the presence of oxygenated organic compounds. Before combustion it is necessary subject them to catalytic hydrogenation (upgrading) which involves, practically, the removal of the oxygen present. Recent studies have evaluated the possibility of using bio-oils for the production of H_2 by catalytic reforming for application in combustible cells [30, 45];
- The aqueous solution: This fraction is derived from the pyrolysis of the feed's humidity. It helps in the dissolution of the organic oxygenated species that originate from the pyrolysis as organic acids, aldehydes, ketones, phenols, which are otherwise difficult to dispose.
- The char: These solid carbonaceous residuals can be used as fuel or find application in the chemical industry.

CHAPTER 5

Environmental aspects

1 Reduction of emissions into the atmosphere

The higher environmental benefit due to the use of biomasses for energy production purposes is connected with the substitution of fossil sources with renewable sources. This is rendered both by the reduction in the use of these sources (that are exhausted because of their nature) and by the decrease in the polluting emissions that are produced from the combustion. The combustion of petroliferous products emits into the atmosphere carbon dioxide that was stored in the vegetable textiles billions of years ago, with an immediate reflex action in terms of an increase in the concentration of greenhouse gases and other pollutants such as nitrogen oxide, sulphur, etc. The combustion of biomasses also generates polluting emissions but they are less dangerous and in smaller quantities compared with the emissions produced by the combustion of fossil fuels. The use of biomasses for energy production purposes represents an instrument to cushion the climate changes and, more generally, to reduce the environmental impacts that are linked to the use of fossil sources.

1.1 The carbon dioxide emissions balance

Biomasses are considered neutral energy sources in terms of greenhouse effects because their combustion does not result in an increase in the concentration of atmospheric carbon dioxide. In fact, given that the quantity of carbon dioxide emitted in the combustion phase is equivalent to that absorbed by the vegetables during their growth, the CO_2 cycle is closed. But taking into account the entire life cycle of the combustibles that form the biomass does not result in a nil balance of CO_2. In fact, the production, working and transport steps often result in negative impacts on the environment that are determined by energy and material consumptions which are necessary to sustain the processes. Table 1 lists the carbon dioxide balance for the fuels obtained from the main biomass production spinnerets. The avoided emissions were estimated as a function of the substituted fossil fuel (coal, diesel, methane) by taking into account the respective calorific powers [2, 3].

Table 1: Avoided and produced carbon dioxide emissions for the main biomass fuel spinnerets [2].

	Avoided emissions	Produced emissions	Medium balance
Wooden biomass	kg CO_2/m^3	kg CO_2/m^3	kg CO_2/m^3
Nippers	450–750	40–55	400–700
Chips	200–350	25–35	170–320
Briquettes and pellet	650–1,100	90–95	560–1,000
Agricultural residuals	kg CO_2/ha	kg CO_2/ha	kg CO_2/ha
Winter-autumn cereal straw	300–1,100	20–75	350–1,050
Stocks, corn cobs, maize sculls	800–1,600	50–110	720–1,520
Rice straw	300–850	25–65	250–800
Culture and fruit arboreal by-products	1,200–6,000	15–60	1,250–5,950
Lignocellulosic from dedicated cultures	kg CO_2/ha	kg CO_2/ha	kg CO_2/ha
Fibre sorghum	22,000–50,000	700–1800	20,000–48000
Kenaf	10,000–35,000	700–1600	9000–34,000
Miscanthus	17,000–58,000	500–1500	16,000–57,000
Common cane	16,000–66,000	500–1500	15,000–65,000
Millet	11,000–50,000	500–1500	10,000–49,000
Poplar	11,000–28,000	500–1500	10,000–27,000
Bio-fuel	kg CO_2/kg bio-fuel	kg CO_2/kg bio-fuel	kg CO_2/kg bio-fuel
Bioethanol from amylaceous	2	0.5–1.1	0.9–1.5
Bioethanol from sacchariferous (beet)	2	0.4–1.1	0.9–1.6
Sunflower bio-diesel	2.7	1.2–1.5	1.2–1.5
Rape bio-diesel	2.7	0.8–2.4	0.3–0.9

1.2 Comparison between the polluting emissions of the main vegetable and fossil origin fuels

1.2.1 Bio-ethanol

The use of bio-ethanol in auto traction, pure or mixed with petrol, improves the quality of the emissions released into the atmosphere from vehicular traffic because it reduces the quantity of the polluting gas emitted. Such improvements can be related to the higher presence of oxygen in the chemical composition of bio-ethanol (equal to 34.7% by weight) compared to gasoline. Oxygen allows the complete combustion fuel, so the quantity of non-combustible compounds is reduced and it results in an increase in the medium life of the engines. Generally, the reduction in the emissions increases as the percentage of bio-ethanol mixed with petrol increases. In some cases, the reduction in the emissions increases in a more than proportional manner; for example, the emissions of carbon monoxide into the atmosphere decrease as the presence of bio-ethanol in the mixture increases. But, this does not happen for the nitrogen oxide (NO_x) emissions which increase as the presence of oxygen in the fuel increases. In fact, it was discovered that the use of bio-ethanol, because of the higher percentage of oxygen, can increase the emissions of certain gasses such as nitrogen oxides and acetaldehyde. A further contribution to the reduction of the dangerous pollutant emissions is due to the use of bio-ethanol as antiknock (ETBE, ethyl *tert*-butyl ether) as a substitute for benzene or MBTE (methyl *tert*-butyl ether). Table 2 shows, for example, the percentage variations in the atmospheric emissions for the proposed use of bio-ethanol in mixtures at 5.5% with the petrol compared to emissions obtained by using gasoline alone. With reference to the improvement of 67% in the acetaldehyde emissions, it is necessary to underline that the presence of this carcinogenic compound is tolerated, even if it is present in high concentrations, because its carcinogenic power is about 10 to 60 times lower than that of the main pollutants (e.g. benzene) which are generated by the combustion of products of fossil origin [2].

Table 2: Percentage variations in some of the polluting emissions for the use of bio-ethanol compared to petrol.

Pollutant	Bio-ethanol in mixture at 5.5% Variation (%)
Carbon monoxide	−10
Volatile organic compounds	−5
Benzene	−25
Butadiene	−11
Formaldehyde	+2
Acetaldehyde	+67

1.2.2 Bio-diesel

Similar to bio-ethanol, the use of bio-diesel in auto traction can also contribute to reducing the environmental impact of vehicles. Bio-diesel, which has an oxygen content (11% by weight) which is higher than that in diesel, is subjected to a better combustion and associated with lower emissions of carbon monoxide, particulate matter and unburned hydrocarbons. By resorting to appropriate modifications in the constructive characteristics of the motors, it is possible to obtain better reductions in the emissions of the main pollutants: direct combustion engines show higher carbon monoxide emissions than the direct injection engines; particulate emissions in the discharge fumes depend on the rotation speed (lower at high regimes). Finally, there are no sulphur oxide emissions, as the raw material, in fact, does not contain sulphur compounds. There are contrasting effects relative to hydrocarbons: the emissions of monocyclic aromatic hydrocarbon, in fact, are reduced (from 20% to 80%), polycyclic aromatic hydrocarbon (up to 98%) and acetaldehyde (from 20% to 40%), whereas the emissions of the short-chain hydrocarbons increase (between 10% and 20%). Nitrogen oxides show variable improvements between 5% and 23% [2, 14]. Table 3 points out the reduction in the emissions of the main pollutants obtained from the combustion of bio-diesel, pure or mixed at 20%, compared to diesel.

Table 3: Percentage variation in some atmospheric emissions compared to diesel [2].

Pollutant	Reduction	
	Bio-diesel at 100%	Bio-diesel at 20%
Carbon monoxide	−42.3%	−12.6%
Particulate	−55.4%	−18%
Burning residual hydrocarbon	−56.3%	−11%

1.2.3 Bio-gas

A further possibility of reducing the atmospheric emissions connected with the use of fossil sources is represented by the use of bio-gas, which is obtained by the anaerobic digestion process, for energy production purposes. Table 4 shows the comparison between the emissions generated by several fuels per energy unit produced.

Table 4: Comparison between the main bio-gas emissions for the main fossil fuels [2].

	SO_2 (kg/TJ)	NO_x (kg/TJ)	Dusts (kg/TJ)
Mineral oils	140	90	20
Gas	3	90	2
Mineral coal	300	150	20
Bio-gas	3	50	3

Bibliography and consulted websites – Part II

[1] ISES ITALIA, *Biomasse per l'energia: guida per progettisti, impiantisti e utilizzatori*, Fondazione IDIS, Città della Scienza, ISES ITALIA, 2004.
[2] Jodice, R. & Tomasinsig, E., Energia dalle biomasse: le tecnologie, i vantaggi per i processi produttivi, i valori economici e ambientali, AREA Science Park: Trieste, 2006, www.area.trieste.it
[3] Bartolazzi, A., *Le energie rinnovabili*, Ulrico Hoepli Editore S.p.a., 2006.
[4] Utilizzo energetico della biomassa, www.aster.it
[5] www.isesitalia.it
[6] FER a biomasse, www.fo.camcom.it
[7] ITABIA (Italian Biomass Association), *Rapporto 2003: Le biomasse per l'energia e l'ambiente*, www.itabia.it
[8] APAT (Agenzia per la protezione dell'ambiente e per i servizi tecnici), Le biomasse legnose: un'indagine sulla potenzialità del settore forestale italiano nell'offerta di fonti di energia, Rome, 2003, www.apat.gov.it
[9] Zilli, M., *Bosco ed energia*, Editori Associati per la Comunicazione: Milan, 2002.
[10] Alberti, M., *La gestione delle ceneri da biomassa: un problema da risolvere*, Progetto Fuoco, Verona, 2002, www.cti2000.it
[11] Righini, W., Indagine conoscitiva sulle caratteristiche di campioni di biomasse legnose prelevate presso l'impianto della società Teleriscaldamento-Cogenerazione Valcamonica, Valtellina, Valchiavenna spa, www.teleriscaldamento.valtline.it
[12] Autori vari, *Biomasse agricole e forestali a uso energetico*, Società A.G.R.A. srl, 2002.
[13] Kavalov, B. & Peteves, S.D., Bioheat Applications in the European Union: An Analysis and Perspective for 2010, http://ecpisystems.com
[14] www.cti2000.it/biodiesel.htm
[15] CTI (Comitato Termotecnico Italiano), Raccomandazione CTI sui biocombustibili solidi: specifiche e classificazione, 2003, www.cti2000.it
[16] www.federlegno.it
[17] Assocarta, Rapporto ambientale dell'industria cartaria italiana 2004, www.assocarta.it
[18] www.liceobisceglie.it
[19] Berton, M., "Il ruolo delle biomasse legnose nella produzione energetica italiana", convegno: "Agroenergie per lo sviluppo rurale", http://venetoagricooltura.regione.veneto.it, 18/9/2006.
[20] www.biomasse.basilicata.it

[21] Jodice, R., atti del convegno *Energia dalle biomasse*, Palmanova 3/5/2006.
[22] Riva, G., *Il pellet: aspetti generali*, Progetto Fuoco, Verona, 2004, www.cti2000.it
[23] www.enerna.com
[24] Keinstar Associates, *I biocombustibili e i biocarburanti*, www.keinstar.it
[25] EurObserv'ER, *dati sulla produzione di biodiesel e bioetanolo in Europa*, 2005, www.energies-renouvelables.org
[26] www.qualenergia.it
[27] European Biodiesel Board, *dati sulla produzione di biodiesel in europa*, www.ebb-eu.org
[28] Pignatelli, V., dossier "Le tecnologie per i biocombustibili e i biocarburanti: opportunità e prospettive per l'Italia", 2006, www.enea.it
[29] www.energie-rinnovabili.net/biofuels
[30] ENEA, "Rapporto energia e ambiente 2003: le fonti rinnovabili", www.enea.it
[31] APAT, "Digestione anaerobica della frazione organica dei rifiuti solidi", 2005, www.apat.gov.it
[32] Piccinini, S., "Biogas: situazione e prospettive", atti del convegno "Bioenergia: un'opportunità da sfruttare per l'agricoltura e l'industria", Piacenza, www.assind.pc.it, 23/2/2006.
[33] Manna, C., "Le fonti rinnovabili 2005: lo sviluppo delle rinnovabili tra necessità e opportunità", ENEA, www.enea.it
[34] www.itabia.it
[35] http://ilsoleascuola.casaccia.enea.it
[36] www2.minambiente.it
[37] "Il divulgatore" n. 12, dicembre 2005, "Energia dal biogas: soluzioni possibili per la zootecnia", www.crpa.it
[38] C.R.P.A. (Centro Ricerche Produzione Animali), *L'integrazione tra digestione anaerobica e compostaggio*, 2006, www.compost.it
[39] ATO-R (Associazione d'Ambito Torinese per il Governo dei Rifiuti), *Il termovalorizzatore della zona nord della provincia di Torino*, www.rifiutilab.it
[40] Materiale didattico del corso *Sistemi di trattamento dei rifiuti* A.A. 2005/2006, Prof. Ing. A. Corti, Facoltà di Ingegneria, Università degli Studi di Firenze, http://didattica.dma.unifi.it
[41] www.fuocolegna.it
[42] www.edilkamin.it
[43] www.forneler.it
[44] "Annesso tecnico n. 2: impianti di incenerimento con recupero di energia", www.atia.it/Citec
[45] ENEA, "Rapporto energia e ambiente 2005", www.enea.it
[46] Materiale didattico del corso di *Interazione tra le macchine e l'ambiente* A.A. 2005/2006, Prof. G. Manfrida, Facoltà di Ingegneria, Università degli Studi di Firenze, http://didattica.dma.unifi.it
[47] www.scuoladottoratoingegneria.unicas.it
[48] http://tecnologie_energetiche.die.unipd.it
[49] www.energetik.leipzig.de

Index

absorber plate, 28–29, 32–35, 39, 41–43, 45, 49, 64
accumulation system, 100, 102
aerobic digestion, 168, 175–176
agricultural compartment, 134, 137
agricultural residuals, 138, 141, 174, 204
agro-alimentary industry, 146, 174
agro-forest behaviour, forest, 135–137
air collector, 51–53
altitude, 6, 11–13, 19, 39
anaerobic digestion, 143–144, 146–147, 162, 165–166, 168, 169–175, 206
Archimedes Project, 93, 110, 115–118

biochemical conversion, 168–182
bio-diesel, 156–159, 181
bioethanol, 204
biogas, 134
biomasses, 203
    classification, definitions, 134–135
    commercial forms, 147–166
        gaseous state combustible biomasses, 162–166
        liquid state combustible biomasses, 147–155
        liquid state fuel biomass, 156–162
    energy from, 167–201
        bio-chemical conversion, 168–182
        energetic biomass conversion, 167–168
        thermochemical conversion, 182–201
    identities, 133–166
    nature, origin, 135–147
        agricultural compartment, 137
        agricultural residuals, 138
        agro-forest behaviour, forest, 135–137
        dedicated cultures, 139–143
        industrial activities, 145–146
        urban residuals, 146–147
        zoo technique compartment, 143–144
brickwork solar walls, 82, 84
briquette, 137, 141, 147, 153–155, 204

carbon dioxide, 9, 27, 162–164, 169, 190–191, 203
cellulose industry, 145
chips, 137, 139, 145, 147–153, 155, 188, 204

chromosphere, 3
codigestion, 173
combustion, 203–205
   direct combustion, 182–190
   commercial forms, 147–166
concentration solar, 89–93, 94, 118
convective envelope, 3
Corona, 3

daily solar radiation, 20
dedicated cultures, 137–143, 159, 160, 176, 180, 204
diagram of solar trajectories, 18–19
direct energy, calculation of, 15–16
direct gain system, 81–82
directed radiation, 9–10, 12

Earth's soil during clear sky days, 9–12
energetic biomass conversion, 167–168
energy from biomass, 167–201
expansion vessel, 70, 73–75
extraterrestrial radiation, 6

firewood, 147
fixed bed gasificators, 191
flat plate collector, 28–39, 54
fluid bed gasificators, 194–197
forced circulation system, 55, 58–60, 62–63

gaseous state combustible biomasses, 162–166
gasification, 182, 190–198
global radiation, 9, 12, 20–21, 36
   instantaneous, 15

heat exchangers, 46, 71–72, 76, 78–80, 109, 172, 182, 189
heliostat, 99–100
Hottel's model, 12
hour angle, 6–7, 14, 16, 20–21
hourly solar radiation, 20–21

inclination, 12, 14–15, 20–23, 42–45, 49, 57
indirect gain system, 81–85
industrial activities, 145–146
infrared emissivity, 33
instantaneous direct radiation, 12–15
instantaneous global radiation, 15
integrated storage collector, 28, 49–51, 60, 62
inversion layer, 3
isolated gain system, 81, 86–87

latitude, 6, 12–13, 22–23, 27, 39, 42, 85
linear parabolic collector system, 95–98, 106
liquid state combustible biomasses, 147–155
liquid state fuel biomass, 156–162
local radiation data retrieval, 21–22

mixed reactors, 172
molten salts
   in parabolic collector systems, 106–117
monthly average solar radiation on inclined surfaces, 19–20
monthly average solar radiation, 19–20

natural circulation system, 55–57, 59, 71, 77, 87
non-return valve, 59, 71–72, 75

oil cultures, 139, 142, 179–180

parabolic dish collector, 95, 104
passive solar heating system, 80–87
pellet, 150–153, 188, 204
photosphere, 3
photovoltaic solar energy, 25
plug-flow reactors, 172–173

pump, 70, 74
  circulation pump, 59, 71
  heat pumps, 60–61
  security group and, 75
pyrolysis, 182–184, 191–193, 198–201

reflection, 9, 15, 33, 89
regulating power unit, 72
restraint valve, 71–72
roof pond, 82, 84–85

scattered radiation, 9
security valve, 70, 73–75
selective surface, 32, 33
solar absorption, 33
solar azimuth, 6–7
solar chimneys, 119–123
solar collector, 18, 26–28, 36, 58, 61, 64, 67, 69, 72, 81, 86, 88–89, 108
  performed by ENEA, 110–111
solar constant, 5–6
solar declination, 7–8
solar greenhouse, 85–86
solar physics, 3–5
solar ponds, 118, 122–126
solar radiation, 3–23, 25, 28, 82, 90, 94, 102
  daily solar radiation, 20
  direct energy, calculation of, 15–16
  on earth's soil during clear sky days, 9–12
  extraterrestrial radiation, 6
  hourly solar radiation, 20–21
  instantaneous direct radiation, 12–15
  instantaneous global radiation, 15
  local radiation data retrieval, 21–22
  monthly average solar radiation, 19–20
solar constant, 5–6
solar physics, 3–5
solar trajectories, diagram of, 18–19
sun in celestial vault, position of, 6–8
true solar time, 16–17
variation in energy, by surfaces' positioning, 22–23
solar tower system, 99
solar trajectories, diagram of, 18–19
spherical collector, 51–52
storage tank, 49–50, 54, 57–59, 63, 75–80, 108
sun in celestial vault, position of, 6–8

temperature sensor, 72–73
thermal solar energy, 25
thermal vector gluid, 30–31, 35, 41, 54, 95–97, 99–101
thermochemical conversion, 182–201
towers, 118–123
trombe's wall, 83
true solar time, 16–17

unglazed collector, 39–40, 53–54, 69
urban residuals, 146–147

vacuum tube collectors, 28, 40–49
variation in energy, by surfaces' positioning, 22–23

water wall, 84–85
wood industry, 145

zenithal angle, 6
zoo technique compartment, 143–144

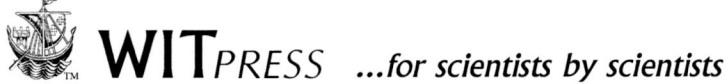

Advanced Computational Methods in Heat Transfer XI

Edited by: **B. SUNDÉN**, Lund University, Sweden, **Ü. MANDER**, University of Tartu, Estonia and **C.A. BREBBIA**, Wessex Institute of Technology, UK

Research and development of computational methods for solving and understanding heat transfer problems continue to be important because heat transfer topics are commonly of a complex nature and different mechanisms such as heat conduction, convection, turbulence, thermal radiation and phase change may occur simultaneously. Typically, applications are found in heat exchangers, gas turbine cooling, turbulent combustions and fires, electronics cooling, melting and solidification. Heat transfer has played a major role in new application fields such as sustainable development and the reduction of greenhouse gases as well as for micro- and nano-scale structures and bio-engineering.

In engineering design and development, reliable and accurate computational methods are required to replace or complement expensive and time-consuming experimental trial and error work. Tremendous advancements have been achieved during recent years due to improved numerical solutions of non-linear partial differential equations and computer developments to achieve efficient and rapid calculations. Nevertheless, to further progress, computational methods will require developments in theoretical and predictive procedures – both basic and innovative – and in applied research. Accurate experimental investigations are needed to validate the numerical calculations.

This book contains papers originally presented at the Eleventh International Conference, arranged into the following topic areas: Natural and Forced Convection and Radiation; Heat Exchanges; Advances in Computational Methods; Heat Recovery; Heat Transfer; Modelling and Experiments; Renewable Energy Systems; Advanced Thermal Materials; Heat Transfer in Porous Media; Multiphase Flow Heat Transfer.

WIT Transactions on Engineering Sciences, Vol 68
ISBN: 978-1-84564-462-8   eISBN: 978-1-84564-463-5
Forthcoming / apx300pp / apx£114.00

*All prices correct at time of going to press but
subject to change.
WIT Press books are available through your
bookseller or direct from the publisher.*